Studies in Space Policy

Volume 30

Series Editor

European Space Policy Institute, Vienna, Austria

The use of outer space is of growing strategic and technological relevance. The development of robotic exploration to distant planets and bodies across the solar system, as well as pioneering human space exploration in earth orbit and of the moon, paved the way for ambitious long-term space exploration. Today, space exploration goes far beyond a merely technological endeavour, as its further development will have a tremendous social, cultural and economic impact. Space activities are entering an era in which contributions of the humanities — history, philosophy, anthropology —, the arts, and the social sciences — political science, economics, law — will become crucial for the future of space exploration. Space policy thus will gain in visibility and relevance. The series Studies in Space Policy shall become the European reference compilation edited by the leading institute in the field, the European Space Policy Institute. It will contain both monographs and collections dealing with their subjects in a transdisciplinary way.

The volumes of the series are single-blind peer-reviewed.

More information about this series at http://www.springer.com/series/8167

Annette Froehlich
Editor

Assessing a Mars Agreement Including Human Settlements

ESPI
European Space Policy Institute

Springer

Editor
Annette Froehlich
European Space Policy Institute
Vienna, Austria

ISSN 1868-5307 ISSN 1868-5315 (electronic)
Studies in Space Policy
ISBN 978-3-030-65012-4 ISBN 978-3-030-65013-1 (eBook)
https://doi.org/10.1007/978-3-030-65013-1

This Springer imprint is published by the registered company Springer Nature Switzerland AG
The registered company address is: Gewerbestrasse 11, 6330 Cham, Switzerland

Preface

The previous worldwide space law essay invitation on the topic of the protection of cultural heritage sites on the Moon was a great success and was well received by the space community. Accordingly, the European Space Policy Institute (ESPI) and the German Aerospace Center (DLR) launched their follow-up invitation for articles by students and young professionals on the topic of assessing a Mars agreement, including human settlement. Excellent contributions were received from around the world. This book thus brings together new and innovative ideas and fresh perspectives on this timely topic that takes place against the backdrop of increasingly widespread discussions on ramping up human activities on the red planet.

According to the European Space Agency (ESA), these activities, currently robotic in nature, but to be followed up by a human presence on Mars, are focused primarily on the search for life, and gaining a greater understanding of the geology, hydrology, atmosphere and broader environmental evolution of the planet. Undoubtedly, with a human presence on Mars (either temporary or permanent), mining and other kinds of resource extraction will also become important activities. Indeed, in recent times various initiatives and techniques for managing a human journey and stay on Mars have been discussed, and it is widely accepted that humans will leave their home planet in the coming decades to become an interplanetary species and to undertake a wide range of activities in space. Undoubtedly, the Moon will be the first target of this expansion. For this reason, the previous worldwide space law essay invitation and resulting book placed its focus on the Moon.

However, as humans ramp up their Martian endeavours, critical legal questions arise there as well. Indeed, one of the central challenges is that the basic legal principles drafted and commonly accepted within the international community were framed by the early space activities of the superpowers and thus took into account the challenges and preoccupations of that time. Therefore, even though the Outer Space Treaty, often called the Magna Carta of international space law, has the scope to cover all space activities in outer space and celestial bodies, it has already been argued to be insufficient in keeping up with the rapid pace of technological progress. The subsequent Moon Treaty expanded on some of the main shortcomings of the Outer Space Treaty, but did not receive widespread acceptance, particularly by the main spacefaring nations. With greater attention now being directed towards human activities and settlement on Mars, the urgency of addressing outstanding legal questions in a manner agreeable to all stakeholders is thus growing.

Accordingly, this book presents a collection of rich contributions from the next generation of space professionals on this topic, assessing various aspects of potential special international rules to be compiled in a future multi-layered Mars agreement. The individual chapters consider not only aspects of regulation—including the precedent of the International Space Station agreement, intellectual property rights, legal considerations around the freedom of exploration and use, and the legal implications of contact with extraterrestrial intelligence—but also address the creation and management of a new society on Mars—including waste streams, criminal justice, human reproduction and childbirth, and the protection of human rights in privately funded settlements.

Given the urgent need to consider these and other international space law issues before actual settlement takes place, and the dearth of publications presenting fresh perspectives and ideas, we stand the best chance of finding agreeable solutions to the multitude of challenges around Mars settlement and exploration by giving a voice to young professionals from around the world. This publication therefore provides a solid contribution to debates that will shape the future of humanity, and helps to ensure that this takes place in the spirit laid down by the Outer Space and other treaties—that the exploration and use of Mars shall be carried out for the benefit and in the interests of all humankind.

October 2020

Dr. Annette Froehlich
European Space Policy Institute (ESPI)
Seconded by German Aerospace Center (DLR)
Vienna, Austria

Contents

Chapter 1
Agreements Protecting Lives on Interplanetary Enterprises (ALIEN): Tailoring a Mars Criminal Justice System

Lauryn Hallet

Abstract The colonisation of Mars has always seemed like a distant dream. However, with the increase of Mars exploration missions and concrete plans from states and private companies to send humans on the Red Planet, the dream may well become reality. The challenges will not merely be technical or scientific, but sociological and legal. Taking into account those last two elements will be decisive in maintaining order amongst Mars settlers and managing criminal activity. To this end, a new criminal justice system tailored to Mars has to be created. On the one hand, it shall be informed by how living on Mars will affect human behaviours and violence. To this end, the study of certain challenging environments and underlying behavioural patterns can be derived to anticipate criminal activity on Mars. On the other hand, questions of jurisdiction to investigate and prosecute those crimes, as well as delegating authority have to be answered and legally framed pre-emptively to inhabiting the planet. However, no existing system is apt to regulate Mars as is. Instead, several cases can be used to build a base for reflection on what elements work and may be reproduced, and what is ineffective and should be discarded.

1.1 Introduction

When one thinks about the colonisation of Mars, they may picture vast deserted red lands with futuristic domes inhabited by selected few humans. A new land harbouring unlimited possibilities. The ultimate break of the last frontier and a chance for the human race to start over. But if humans are to populate Mars with their knowledge, dreams and good intentions, an uninvited guest may come along; crimes. It has been asserted that criminal activity on Mars settlements is practically impossible in reason of the status of astronauts.[1] The idea that persons of such pedigree would be impermeable to violence because of their intrinsic respect for discipline and the

[1] Christopher J. Newman, 'Exploring the problems of criminal justice in space' (2016) 2(8) *ROOM*.

L. Hallet (✉)
University of Bristol, Bristol, UK
e-mail: laurynhall@hotmail.com

A. Froehlich (ed.), *Assessing a Mars Agreement Including Human Settlements*, Studies in Space Policy 30, https://doi.org/10.1007/978-3-030-65013-1_1

fact that they have been selected through a precise methodology, is greatly mistaken and should be addressed and debunked.[2] Violence is not a characteristic specific to a low socioeconomic status. According to studies made by the American Psychological Association, "exposure to violence is not bound to age, socioeconomic status, level of education, or occupation".[3] Thus, no choice of a particular age group, title, or level of education will a priori prevent an individual from committing a crime. As the saying goes "it is not a matter of if, but when". It is no longer the time to deny the likelihood of criminal behaviour within Mars settlements, but rather to lay probable scenarios based on relevant experiences on Earth and to frame criminal justice accordingly.

To this end, it is presently argued that no criminal justice models from Earth could be applied on Mars as is. Mars, because of the new society it will host and the extreme peculiarity of the context, has to be given a tailored legal framework which should be informed by a deep understanding of human nature and psychology. To this end, the first section looks at certain settings such as the International Space Station (ISS), bases in Antarctica, submarines, and isolated communities which can offer a preliminary vision on how difficult environments impact human behaviour and its propensity to violence.

Asserting the probability of Martian crimes and the ensuing necessity for a Mars criminal justice system is merely the beginning. If no existing justice system is apt to frame and regulate criminal activities on Mars, there are nonetheless several models to be used as guides. The second section looks at how state jurisdiction to prosecute and investigate is asserted in space, in Antarctica, on the high seas, and in the Indian Country. Each case can incite a reflection as much on the useful elements to be transferred to the Mars system, than on the elements to avoid reproducing.

1.2 Criminal Activity Anticipation Through Observable Human Behaviour

This article seeks to study violent crimes with their *locus* on Mars, in which both parties involved are settlers. Interplanetary crimes, such as those that could occur from a cyberattack from one planet onto the other are not reviewed in the present. Violent crimes mean *inter alia* murder, assault, sexual assault, or any other act of a nature such as to damage the physical or psychological integrity of the victim.

As advanced by Goetz, violence and humans have always been and always will be coexisting.[4] Against this backdrop, scientific research has sought to find patterns within our history of violence. Aside the chances of settlers inadvertently hosting

---

[2] Ibid.

[3] American Psychology Association, 'Violence and socioeconomic status' Factsheet. https://www.apa.org/pi/ses/resources/publications/factsheet-violence.pdf.

[4] Aaron T. Goetz, 'The Evolutionary Psychology of Violence', (2010) *Psicothema* 22(1), 15–21.

psychopaths and sadists amongst the settlers, violence on Mars will arise from situations that can be to a certain extent anticipated.[5] The reason is that violence observed on Earth is not incumbent on location but on circumstances and human nature.[6] Research have aimed to retrace how humans have historically answered with violence in response to what was called "adaptive problems" such as stealing, self-defence, infidelity, deterrence of future aggression, challenge to status and hierarchy.[7] Those are all instances that could happen on the settlements.

In addition to violence in "standard" settings, highly challenging environments produce intrinsic triggers to violence. Such environments can help anticipate the violent behaviours to be sparked from living on Mars settlements. Those are critical elements to consider in order to create an adequate Mars criminal justice system.

1.2.1 Human Behaviour in Space or Space-Simulated Expeditions

The sole human-populated permanent habitat in space is the ISS. There, astronauts from different countries and a restricted pool of nationalities cohabitate in groups of six. They are highly monitored in real-time and follow a precise and full schedule. Their missions may last from a few days to a few months. They do not have to function as a society and their purpose is the advancement of space exploration and space activities in general. Once their contribution to the scientific mission is done, they come back on Earth.

There is a theoretical resemblance between life on the ISS and life on Mars settlements, and a prevailing assumption is that the former is the optimal example to predict the latter. This article contends the opposite. First, the social dynamics are entirely different. The ISS does not contain a sufficient number of people to recreate a social structure as will be done when populating Mars. A growing community implies progressively increasing social relations, with the increasing potentiality to interact with disliked people one has not chosen to live with, as well as drastic and irrevocable changes in society's structure. Second, the time of the expedition is short compared to the one-way trip that will be presented to settlers. There will be no prospect of coming back to Earth to families and to "normal life" as there is for astronauts. Third, the ISS is highly monitored and astronauts are certain that any wrongful act will lead to investigation and prosecution when back on Earth. On Mars, contacts with Earth would not be in real time. Unless a well-rounded Mars criminal justice system is erected to provide adjusted deterrent and punitive measures, some could be impelled to commit wrongful acts. On the one hand, the prospects of being brought to justice on Earth is almost null, and on the other hand, there would be an absence of tangible repercussions.

---

[5] Aaron T. Goetz, 'The Evolutionary Psychology of Violence', (2010) *Psicothema* 22(1), 16.

[6] Aaron T. Goetz, 'The Evolutionary Psychology of Violence', (2010) *Psicothema* 22(1), 17.

[7] Ibid.

Moreover, NASA has sought to study human behaviour in order to prepare for Mars through a series of simulation experiments, called HI-SEAS. Supposed to carry research on sleep, fatigue, mood, and stress, the final mission was prematurely stopped on day four, after one subject incurred an electric shock and needed proper medical assistance. If research were cancelled, lessons can still be learned from the failures of the experience. Once on Mars, it will not be possible to pull off the plug when something goes wrong. On HI-SEAS, although the crew received first aid training, they could not help their crew-mate. Also, they knew assistance was on its way and still panicked. On Mars, if help cannot be provided by the settlers, they will have to take decisions and bear the consequences. Feeling helpless, responsible for the loss of someone, disagreeing with the group decision, those are factors that can create tensions and violent reactions. Furthermore, a maximum of six individuals were selected per each experiment.[8] When the schedule went as planned it lasted from four months to a year.[9] Similarly to the ISS argument, those shorter-term experiments can only offer a partial view on the psychological consequences of a one-way trip to Mars and its effects on stress levels and violence.

1.2.2 Human Behaviour on Antarctica Scientific Expeditions

If the landscape in Antarctica appears immaculate and serene, research facilities have not been exempted from criminal activity. In 1996, a man attacked his crew-mate with a hammer at McMurdo Research Station (US) which led the FBI to investigate the case *in-situ*.[10] Four years later, an Australian died of methanol poisoning at the Amundsen-Scott South Pole Station (US) situated on the Ross Dependency (New Zealand territory). Although New Zealand proceeded to protracted investigations and hearings to determine the causes of his suspicious death, and without much cooperation from the US Government, no conclusive evidence was found and the fatality was not considered a homicide.[11] McElrea, coroner[12] on the case said the disorganisation of the investigation revealed the need to establish a set of clear criminal justice rules

[8]University of Hawaii, 'HI-SEAS Hawaii Space Exploration Analog and Simulation' Media Kit (2018). https://hi-seas.org/wp-content/uploads/2018/02/HI-SEAS-MediaKit_Jan2018.pdf.

[9]Ibid.

[10]Mindy Weisberger, 'No Evidence Russian Engineer Stabbed Antarctica Colleague for Spoiling Book Endings' (2018) *Livescience*. https://www.livescience.com/64012-antarctica-stabbing.html.

[11]Ibid.; Michele Debczak, 'Death at the South Pole: The Mystery of Antarctica's Unsolved Poisoning Case' (2019) *Mentalfloss*. https://www.mentalfloss.com/article/579732/mysterious-death-rodney-marks-scientist-who-was-poisoned-antarctica.

[12]A coroner is an "official who examines the reasons for a person's death, especially if it was violent or unexpected". Cambridge Dictionary. https://dictionary.cambridge.org/dictionary/english/coroner.

in Antarctica.[13] Most recently in 2018, the Bellingshausen Station (Russia) was the theatre of an attempted murder when a man stabbed his crew-mate on multiple occasions during an alleged mental breakdown.[14] The perpetrator of Russian nationality, surrendered to the director of the station and was detained in an Antarctica church for a week as there is no police or proper means of detention.[15] He was then handed out to the Russian police upon return on Russian soil.

Several factors could explain instances of violence on the continent in spite of the selection process for researchers. Professor of Psychology Peter Suedfeld, expert in Restricted Environmental Stimulation Therapy, dedicated research to coping mechanisms and adaptation to extreme and stressful situations such as those provoked by space or Polar Regions.[16] He confirms that people act differently when subjected to stress and when not, *e.g.* someone usually friendly in a comfortable environment can become hostile and egotistic in challenging conditions.[17] He also evoked the tight living space leaving little room to privacy, limited contact with the outside world, the impossibility to go outside at times, the perspective of being stuck with unchosen individuals, distance from home; those are all element incontestably affecting even the strongest and most prepared of minds.[18]

The behaviours triggered by challenging environments come in addition to the set of predictable violent situations listed *supra* and exacerbate initial propensions to violence. Recreating unpleasant and extreme situations is a good start to study coping mechanisms but it is not enough.[19] Different scenarios cause different types of stress.[20] There is however one constant studied in Antarctica and with which settlers will also have to compose. On top of boredom and claustrophobia, life conditions in Antarctica have aggravating effects on sleep patterns; disrupted in reason of either permanent daylight or permanent darkness.[21] Suedfeld has affirmed that "people

[13] Michele Debczak, 'Death at the South Pole: The Mystery of Antarctica's Unsolved Poisoning Case' (2019) *Mentalfloss*. https://www.mentalfloss.com/article/579732/mysterious-death-rodney-marks-scientist-who-was-poisoned-antarctica.

[14] Ibid.

[15] Mindy Weisberger, 'No Evidence Russian Engineer Stabbed Antarctica Colleague for Spoiling Book Endings' (2018) *Livescience*. https://www.livescience.com/64012-antarctica-stabbing.html.

[16] Rebecca McPhee, 'Coping with Isolation and Extremes: Interview with Peter Suedfeld' (2020) *Explorersweb*. https://explorersweb.com/2020/06/17/coping-with-isolation-and-extremes-interview-with-peter-suedfeld/.

[17] Ibid.

[18] Michele Debczak, 'Death at the South Pole: The Mystery of Antarctica's Unsolved Poisoning Case' (2019) *Mentalfloss*. https://www.mentalfloss.com/article/579732/mysterious-death-rodney-marks-scientist-who-was-poisoned-antarctica.

[19] Rebecca McPhee, 'Coping with Isolation and Extremes: Interview with Peter Suedfeld' (2020) *Explorersweb*. https://explorersweb.com/2020/06/17/coping-with-isolation-and-extremes-interview-with-peter-suedfeld/.

[20] Ibid.

[21] Michele Debczak, 'Death at the South Pole: The Mystery of Antarctica's Unsolved Poisoning Case' (2019) *Mentalfloss*. https://www.mentalfloss.com/article/579732/mysterious-death-rodney-marks-scientist-who-was-poisoned-antarctica.

with disrupted circadian cycles are more likely to exhibit aggressivity, irritability, sensitivity and to misinterpret" social interactions.[22]

1.2.3 Human Behaviour in Isolated and Tribal Communities

There will be no community more remote and isolated that the one of Mars settlers. As such, lessons can be learned from the effects of isolation and of the social particularities of small communities and tribes on violence and crime.

In the Indian Country, according to Tribal leaders, Indian Reservations are seeing increasing cases of violent crimes responsible for the demise of the community and affecting it to a higher extent when compared against the national statistics.[23]

Another unfortunate specificity of small and isolated communities is the higher propensity to commit gender crimes. There is statistical correlation between geographic isolation and higher occurrences of domestic violence in which the isolation factor is used by the perpetrator as a way to control the victim.[24] According to a study carried within the Australian Institute of Family Studies, woman in rural and remote areas experience more domestic violence than woman in urban areas and "face specific issues because of the geographical and social characteristics of living in small communities".[25]

The majority of Indian Native women are inflicted violence, sexual violence, and death in rates defying national averages. It has been reported that "in some reservations, indigenous women are murdered at more than ten times the national average".[26]

Attempting to understand the reason behind the patterns of violence in small communities, it has been advanced that "small towns have populations large enough to have the difficult problems of a big city, while at the same time are hit the hardest economically, so they don't have specialized services and policing needed to deal with family violence."[27] While not directly due to economic factors, a settlement on Mars will have social similarities with isolated and tribal communities and runs the

---

[22]Ibid.

[23]US Department of the Interior, Indian Affairs, 'Tribal Court System'. https://www.bia.gov/CFRCourts/tribal-justice-support-directorate.

[24]Sara Zaske, 'Small towns have highest risk of intimate partner violence' (2020) *Washington State University Insider*. https://news.wsu.edu/2020/08/06/small-towns-highest-risk-intimate-partner-violence/; Monica Campo and Sarah Tayton, 'Domestic and family violence in regional, rural and remote communities' (2015), *Australian Institute of Family Studies*. https://aifs.gov.au/cfca/publications/domestic-and-family-violence-regional-rural-and-remote-communities.

[25]Ibid.

[26]Indian Law Resource Center, 'Safe Women, Strong Nations'. https://indianlaw.org/safewomen.

[27]Sara Zaske, 'Small towns have highest risk of intimate partner violence' (2020) *Washington State University Insider*. https://news.wsu.edu/2020/08/06/small-towns-highest-risk-intimate-partner-violence/.

risk of incurring the same issues and desperate needs of police and other relevant services.

1.2.4 Human Behaviour on Submarines

Sexual violence in the military is a recurrent problem, and on submarines the consequences may be more severe.[28] In addition to the common fact that the aggressor will be known amongst a bonded group as a colleague or a friend impacting the victim's decision to report, movements on a submarine are highly restricted and there is no way to escape from an assault, or take distance from the aggressor. This may instill fear of reaction for the victim. If they find the courage to do so, the current system is such that they may still not be heard, and justice may not be rendered. In the nine years during which women have served on US submarines, multiple cases of sexual violence have erupted (knowing that many more go unreported).[29] In 2019, it was revealed that "male sailors created rape lists of female shipmates" on a US submarine.[30] The incident took a year to be heard after the first reports had been made.[31] The infectivity of the system only adds to the victim's sentiment that they will not be heard and that the perpetrator will go unpunished. Implicitly inhibiting the victim's will to report contributes to maintaining an environment that already fosters violence and crime.

Constrained environment with claustrophobic living spaces and small groups of people to cohabit might not only exacerbate violence on the perpetrator's side, it also tends to reinforce the vulnerability and isolation of the victim. This has to be considered in preparing for Mars. A criminal justice system needs to address gender crimes and sexual violence.

1.3 State Jurisdiction and Delegation of Authority

The probability of crimes being perpetrated by settlers has been assessed and attention has been brought on certain typologies. This preliminary assessment should be complemented by clear rules as to which state has jurisdiction to investigate and

---

[28] Jim Garamone, 'Defense Officials Tout Progress in Fight Against Sexual Assault' (2020) *DOD News*. https://www.defense.gov/Explore/News/Article/Article/2172056/defense-officials-tout-progress-in-fight-against-sexual-assault/; Serene Fang, 'In the same boat: Navy integrates women aboard submarines' (2014) Aljazeera America. https://america.aljazeera.com/watch/shows/america-tonight/articles/2014/11/2/in-the-same-boatnavyintegrateswomenaboardsubmarines.html.

[29] Mark D Faram, 'Male sailors created 'rape lists' of female shipmates on Georgia-based submarine' (2019) *NavyTimes*. https://www.navytimes.com/news/your-navy/2019/05/19/male-sailors-created-rape-lists-of-female-shipmates-on-georgia-based-submarine/.

[30] Ibid.

[31] Ibid.

prosecute crimes on Mars, and who exercises authoritative functions. To adequately determine jurisdiction on Mars and delegate authority to the right body that will maintain discipline on a daily basis, some existing models can be useful.

1.3.1 State Jurisdiction and Delegation of Authority in Space

According to general Space Law and article VIII of the OST, jurisdiction is given to the country to which the object belongs, *i.e.* the country where the object is registered. This is an application of the quasi-territorial jurisdiction.

On board of the ISS, Article 22 of IGA (Inter-Governmental Agreement) providing for criminal jurisdiction complemented by the ISS Crew Code of Conduct, give priority to the active personality principle where jurisdiction is given to the country of the perpetrator.[32] For instance, when a US NASA astronaut was accused of hacking the bank account of their former spouse from a NASA computer whilst living aboard the ISS, the case was investigated by the US and treated under US law as a purely domestic legal affair.[33]

Also, IGA grant the commander a certain level of authority.[34] Precisely, a disciplinary authority is attributed to the space object commander, and the state's jurisdiction is applied through that person.[35] The Crew Code of Conduct established the commander as the ultimate authority and puts in their hands the safety of crew and the insurance of order and discipline with the authorisation to use proportional and necessary means to achieve such a mission.[36]

It has been asserted that the ISS framework would be the best contender to apply in an interplanetary mission such as the human exploration of Mars; that the active personality principle and the nature of the commander's authority would be the best option.[37] However, both are rather ill-fitted. First, giving jurisdiction to the country of the perpetrator is not desirable. What if there is no known perpetrator. Also, how

---

[32] Michael Chatzipanagiotis, Rafael Moro-Aguilar, 'Criminal Jurisdiction in International Space Law: Future Challenges in View of the ISS IGA' (2014) 57th IISL Colloquium on the Law of Outer Space, 1.

[33] Chelsea Gohd, Who Investigates a Crime in Space? 29 August 2019, Space.com, https://www.space.com/who-investigates-space-crime.html.

[34] Michael Chatzipanagiotis, Rafael Moro-Aguilar, 'Criminal Jurisdiction in International Space Law: Future Challenges in View of the ISS IGA' (2014) 57th IISL Colloquium on the Law of Outer Space, 1.

[35] Michael Chatzipanagiotis, Rafael Moro-Aguilar, 'Criminal Jurisdiction in International Space Law: Future Challenges in View of the ISS IGA' (2014) 57th IISL Colloquium on the Law of Outer Space, 5.

[36] Michael Chatzipanagiotis, Rafael Moro-Aguilar, 'Criminal Jurisdiction in International Space Law: Future Challenges in View of the ISS IGA' (2014) 57th IISL Colloquium on the Law of Outer Space, 9.

[37] Michael Chatzipanagiotis, Rafael Moro-Aguilar, 'Criminal Jurisdiction in International Space Law: Future Challenges in View of the ISS IGA' (2014) 57th IISL Colloquium on the Law of Outer Space, 11.

can the state *de facto* investigate and prosecute its national from such a distance, especially if it has no other nationals on Mars. In reality, it goes further, how can any state pretend to have jurisdiction over the investigation and prosecution when it was not present during the act and will never be able to send a team on site or bring the perpetrator in front of its courts? Assertion of state jurisdiction is untenable in the Mars context. The Mars criminal justice system will need to integrate a supranational body which will have jurisdiction over criminal cases.

Second, the problem with giving disciplinary authority to the commander of the mission especially arises when different commanders from different missions will have their authority opposed and challenged by one another. This creates a conflict of overlapping authorities but may also trigger the "challenge to status and hierarchy" issue described *supra*. The latter cannot be taken lightly. Indeed, a research on homicide and status competition studied the psychology of status hierarchy negotiation and status maintenance. It concluded that the latter "may lead to antisocial behaviour when the perception of status-inappropriate behaviour by each individual engenders a status protecting and promoting arms race" that ends in violence.[38] Again, it is important that only one person or one organised body of persons answering to the same overarching authority are competent to maintain law and order on the settlements.

1.3.2 State Jurisdiction and Delegation of Authority in Antarctica

Article VIII of the Antarctica Treaty provides that observers, scientific personnel exchanged between stations, and staff accompanying them, shall only be subject to the jurisdiction of their state.[39] The active personality principle ensuing from this does not *ergo* applies to individuals not covered by Article VIII, *e.g.* military personnel and most scientific and support staff.[40]

In the Australian poisoning case mentioned above, the quasi-territorial principle was applied. Active personality could not have been applied as there was no perpetrator, or at least the New-Zealand investigation determined there was none. In the two other incidents mentioned, the Russian national was subjected to the jurisdiction of Russia, and the American cook, to the jurisdiction of the US. Those occurrences raised a significant flaw in the system that applied. There is a lack of coordinated efforts to lead one single investigation. Rather, concurrent investigations have been conducted by distinct entities which may had different interests. The comment from McElrea to establish a clear set of criminal rules to avoid overlapping or confusing attributions of jurisdiction also reveal the type of legal conundrums that will hamper

[38] Aaron T. Goetz, 'The Evolutionary Psychology of Violence', (2010) *Psicothema* 22(1), 17.

[39] Antarctic Treaty (entered into force 23 June1961) 402 UNTS 71, Art VIII.

[40] Elizabeth K. Hook, 'Criminal Jurisdiction in Antarctica' (1978) *University of Miami Law Review* 33(2), 492.

the rendering of justice on Mars. Jurisdiction should be attributed to a sole and clear figure of justice in situ.

With respect to delegating authority to an individual, Article VII of the Antarctica Treaty provides states with the ability to appoint an observer of the nationality of that state in order to carry inspections and do so in complete freedom of access at any time and in any areas of Antarctica (all stations, all installations, etc.).

The appointment of an observer to carry inspections could be used on Mars to a certain degree. In the hypothesis where a crime occurs, an inspector could be appointed to carry an investigation. However, it should not be the case that a state appoints one of its national. As stated, if the settlements do not count another individual of the corresponding nationality, the system fails. It should not be up to a specific country but rather to a supranational body to elect the authority.

A major weakness in the management of crimes in Antarctica is that there is no proper policing authority and suitable means of containment for wrongdoers. On this, careful attention should be paid to the human rights protecting a person under detention or imprisonment. In absence of proper means of containment on Mars and clear provisions as to whom may proceed to the detention or arrest and under what circumstances, arbitrary arrest and detention, and ill treatments could lead to breaches of Human Rights Law.

1.3.3 State Jurisdiction and Delegation of Authority on the High Seas

The high seas, like outer space, are territories outside of national jurisdiction. On the high seas, the general rule is the exclusive authority of the flag state.[41] The flag state's jurisdiction and domestic law are to be exercised on both the ship and the crew in regard of administrative, technical, and social matters.[42] The UN Convention on the Law of the Sea (UNCLOS) allows flag states to conduct inquiries into "casualties or incident on the high seas when vessels flying their flag cause loss of life to or seriously injure foreign individuals".[43] The states involved are to cooperate during the investigation.[44]

One notable exception to the quasi sacrosanct jurisdiction of the flag state is piracy. In that case, authority of the flag state was rather obstructive, especially when pirates use flags of convenience and the flag states have no interest in taking measures. In an attempt to circumvent previous legal loopholes, UNCLOS compelled all states to cooperate in the "repression of piracy on the high seas or in any other place outside

[41]United Nations Convention on the Law of the Sea (UNCLOS) (entered into force 16 November 1994) 1833 UNTS 3, art 94.

[42]UNCLOS, art 94.

[43]UNCLOS, art 94(7).

[44]Ibid.

the jurisdiction of any State".[45] Further, any state may board and seize pirate vessels regardless of their flag, arrest the individuals on board, and have its courts impose penalties, in other words acts of piracy are subjected to universal jurisdiction.[46]

Authority on the high seas is placed in the hands of the ship master. The exact extent of the authority and collateral responsibilities depend on the domestic laws of the flag state.[47] Generally, significant discretion is left to the ship master because of the special and dangerous nature of sailing on the high seas.[48] Their main duty is to ensure the "safe operation of the vessel and the protection of its contents".[49] Stemming from that is the duty to ensure safety of the people and cargo within the ship, to the effect of which they hold the power to keep order on the ship.[50] The translation is twofold. First, there is a disciplinary plane on which the master may either take reasonable measures in response to an act or in deterrence of future acts.[51] Assessment of necessity and proportionality has to be measured against the nature of the wrongdoing.[52] The master may proceed to the arrest of a person when the order or safety of the ship are jeopardised, or they can even use force in cases of mutiny for instance.[53] Second, the master may act from a criminal law enforcement plane, and after having taken the necessary and reasonable disciplinary measures, write a report to relevant superior authorities.[54]

The quasi-territorial principle has been retained on the high seas as flag states have full authority. The reason why applying this principle on Mars is problematic has been established in the previous sections. Nonetheless, what is interesting from the high seas regime is the departure from a state-centric form of jurisdiction in the case of piracy.[55] Piracy engenders problems the system of the time was not adequately managing and as such universal jurisdiction was seen as a way to detangle the jurisdictional conundrum. On Mars, universal jurisdiction over crimes may still not be the best option, as it ultimately relies on the actions of individual states. However, the instance of piracy does demonstrate that principles of jurisdiction do not absolutely have to be state-centric. A supranational body is needed and it should be tailored to answer the legal loopholes or jurisdictional issues that would arise from crimes committed on settlements.

---

[45] UNCLOS, art 100.

[46] UNCLOS, arts 105, 110.

[47] John AC Cartner et al., *The International Law of the Shipmaster* (Informa 2009), 88.

[48] Ibid.

[49] John AC Cartner et al., *The International Law of the Shipmaster* (Informa 2009), 145.

[50] John AC Cartner et al., *The International Law of the Shipmaster* (Informa 2009), 150.

[51] Ibid.

[52] John AC Cartner et al., *The International Law of the Shipmaster* (Informa 2009), 151.

[53] Ibid.

[54] Ibid.

[55] Mark Chadwick, 'Emerging Voices: Theorising Universal Jurisdiction–Time to Reappraise the "Piracy Analogy"?' (2019) *OpinioJuris*. https://opiniojuris.org/2019/08/19/emerging-voices-theorising-universal-jurisdiction-time-to-reappraise-the-piracy-analogy/.

1.3.4 State Jurisdiction and Delegation of Authority in the Indian Country

The jurisdictional and policing system set in the Indian Country deserves attention.[56]

US Tribes have a particular legal and political status in that they are deemed as sovereign nations within the US. Tribes have immunities, rights, and obligations. For instance, the Indian Country has jurisdictional rights over matters involving Indian Natives. With respect to crimes, the Indian Country has competence over misdemeanours involving Indians and happening within tribal jurisdiction, the latter referring to the American Indian Reserve Land. Sentences are delivered according to the laws and customs of the Tribe, but imprisonment and fine sentences are systematically constrained to a maximum given time period and amount by the US Congress and the Tribal Law and Order Act of 2010. More serious crimes, *i.e.* felonies which are also crimes under Federal Law, must be heard before a Federal Court, even if they involve Indians within Indian Country.

A jurisdictional issue has been revealed in places where tribal and state lands are contiguous. It is not always clear whether a tribal officer could have jurisdiction over non-Natives.

Concerning authority, the Indian Country has formed three types of law enforcement bodies. The first, the BIA Police (Bureau of Indian Affairs), provides services to tribes with no police body of their own in place.[57] The second is the Tribal Police for which either officers are funded through self-determination and self-governance laws or can be fully funded by tribes, in which case they gain complete law enforcement jurisdiction over tribal lands.[58] The third body is the local non-Indian police operating when the tribal land is within a political jurisdiction.[59] In reason of this tripartite superposition of jurisdiction and the geographic isolation of some tribes, law enforcement on those lands may reveal both confusing and arduous.[60]

Although the jurisdiction and authority system in the Indian Country is intricate because of its interrelation with the US federal system and the blurry line between the two, it highlights the necessity for certain groups of people to have certain rights to render justice. In this case, the context and history are such that the Indian Country is best suited to optimally manage actions of Natives on Reserve Land. Similarly, Mars settlers although still being linked with their federal country of origin, will live on a land that has its own particularities, deserving its own set of rules. Moreover, a police body, similar to one of the models operating on Reserve Land, should exist alongside or under the proposed supranational body. This option should certainly be

[56]Several other models exist, *e.g.* in Australia, Canada, and New Zealand. However, for the sake of conciseness, this article focuses on the Indian Country in the US.

[57]John Kiedrowski, Trends in Indigenous Policing Models: An International Comparison (2013) *Public Safety Canada*, 16.

[58]Ibid.

[59]Ibid.

[60]John Kiedrowski, Trends in Indigenous Policing Models: An International Comparison (2013) *Public Safety Canada*, 17.

favoured over the appointment of individual disciplinary figures as was the case on the ISS, Antarctica, and on the high seas.

1.4 Conclusion

This article seeks on the one hand, to establish the need for a criminal justice system on Mars by selecting several scenarios and studying their impact on human behaviour. Certain patterns are noticed from those behaviours and they can contribute to predict criminal activity on Mars to an extent. On the other hand, the article aims to provide a preliminary reflection on some essential elements that will form the said system. Those elements can come from obvious environments such as space and the ISS, but other less evident example should not be set aside. That is the case for jurisdiction and authority in Antarctica and on the high seas. Moreover, Tribes and their ways may offer an unusual but constructive perspective how to think outside of the box when planning for the organisation of Mars settlements and catering to needs and actions of their inhabitants.

Ultimately, many more questions are still to be asked and answered. Whether it is about models of governance and self-governance, statehood, self-determination, immunities, separation of powers, organisation of courts, detention facilities; those are all topics that have to be comprehensively studied and incorporated in the bigger picture of not just a criminal justice system on Mars, but a general political and justice system so that Mars settlers can thrive as a new society living in extraordinary settings with extraordinary and unique implications.

Lauryn Hallet holds an International Law LLM from the University of Bristol, currently interning as a space policy researcher. Her fields of interest, whether professional or personal, are space law and policy, maritime security, languages and cosmology.

Chapter 2
Is Human Settlement on Mars Marking the Beginning of a New Era of Intellectual Property Rights Protection in Outer Space?

Terezie Němcová

Abstract Since commercial players continuously keep joining activities in outer space, the interest in protecting their ideas in form of intellectual property in these activities might become an urging topic, especially once humans settle on a new territory as Mars. Despite the fact that several public international agreements and treaties related to outer space activities as well as treaties related to intellectual property rights entered into force decades ago, an international regulatory framework for intellectual property rights protection either on Mars or elsewhere in outer space still is non-existent and possibly preventing or limiting newcomers from the private sector to participate in outer space activities together with governmental space agencies. This essay provides an overview on the already existing treaties and agreements regarding outer space as well as public international agreements on intellectual property rights and their scope. It further provides an outline and comparison of the possible future scenarios of intellectual property rights protection on the Red Planet and highlights their most effective and convenient aspects relevant for the international community involved in outer space activities. This essay concludes in the illustration of the possible ways of implementing these solutions to a future Mars Agreement.

2.1 Introduction

Since the beginning of the space age in 1969 on the Moon, the major players as well as the programmes and plans have changed.[1] The Moon Agreement which was established for the purposes of the national agencies conquering the Moon still is in

[1]Neel V. Patel, *'Are We in the Middle of a New Space Race for This Century?'* (*MIT Technology Review*, 7 February 2020). www.technologyreview.com/2020/02/07/349033/are-we-in-the-middle-of-a-new-space-race-for-this-century/. Accessed 13 September 2020.

T. Němcová (✉)
Faculty of Business, Economics, and Law, Friedrich-Alexander-Universität Erlangen-Nürnberg (University of Erlangen-Nuremberg), Erlangen, Germany
e-mail: terezienemcova@gmail.com

A. Froehlich (ed.), *Assessing a Mars Agreement Including Human Settlements*, Studies in Space Policy 30, https://doi.org/10.1007/978-3-030-65013-1_2

force today, together with the other four main treaties and agreements which shape the area of Space Law.[2] Even though the space race was "the fruit of intellectual creations", the protection of intellectual property rights (IPR) is still only limited in outer space.[3] This has the consequence that parties which are cooperating and using intellectual property in outer space activities have to enter into international cooperation agreements.[4] As more commercial actors are joining the governmental space agencies in outer space activities, the interest in protection of IPR and therefore securing of the investments made for their inventions rises.[5] Space agencies have already started operating on Mars,[6] the private sector, however, also has great aspirations in this regard.[7] For this reason, the establishment of a Mars Agreement generally seems to be inevitable.

Despite the fact that both of these Agreements would regulate a specific regime on a celestial body, their genesis would very probably differ. The Moon Agreement arose from the negotiations undertook in late 70s in the context of the space race and established regulations of the activities of human kind in outer space, in order to maintain a safe and fair regime on the Moon.[8] It is probably for the historical background that the Moon Agreement does not include any details on the protection of IPR, as the states participating in space activities at that time were concerned about other issues such as the use of nuclear power and other safety aspects having priority for regulatory measures and were merely the reasons for entering into international agreements.[9] The space race, at least at the very beginning and therefore at the times of the genesis of the Moon Agreement, was dominated by state related space agencies which did not have to rely on the recovery of the investments made for the costly

---

[2]Issue Paper on intellectual property in context of activities in outer space prepared by the International Bureau of WIPO, '*Intellectual Property and Space Activities*' (April 2004). www.wipo.int/export/sites/www/patent-law/en/developments/pdf/ip_space.pdf. Accessed 14 September 2020, 7–8, mn 29–34 [hereinafter as *IP WIPO*].

[3]WIPO, '*Patent Experts Issues: Inventions in Space*'. www.wipo.int/patents/en/topics/outer_space.html. Accessed 13 September 2020.

[4]*Id*, 4–5, mn 19; Michael Gerhard, '*Article VI*' in Stephan Hobe, Bernhard Schmidt-Tedd and Kai-Uwe Schrogl (eds), *Cologne Commentary on Space Law* (Volume I, Carl Heymanns Verlag 2009), 121–122, mn. 74.

[5]Andy Davies, '*Patenting Inventions in Space*' (*Reddie & Grose*, 3 January 2019). www.reddie.co.uk/2019/01/03/patenting-inventions-in-space/. Accessed 13 September 2020; Stefan Paterson, Robert Wulff, '*The Role of Intellectual Property in Space*' (*Spacetech Asia*, 31 July 2018). www.spacetechasia.com/the-role-of-intellectual-property-in-space/. Accessed 13 September 2020.

[6]ESA, '*Mars Express*'. www.esa.int/Science_Exploration/Space_Science/Mars_Express. Accessed 14 September 2020; NASA, '*Mars Today: Robotic Exploration*'. www.nasa.gov/mission_pages/mars/main/index.html. Accessed 14 September 2020.

[7]SpaceX, '*Mars and Beyond*'. www.spacex.com/human-spaceflight/mars/. Accessed 14 September 2020.

[8]Stephan Hobe, Peter Stubbe, Fabio Tronchetti, '*Historical Background and Context MOON*' in Stephan Hobe, Bernhard Schmidt-Tedd and Kai-Uwe Schrogl (eds), *Cologne Commentary on Space Law* (Volume II, Carl Heymanns Verlag 2013), 336–338, mn. 1, 6.

[9]*Id*, 338, mn. 5.

space related projects as much as the actors from the commercial sector do today.[10] However, as stressed by the World Intellectual Property Organisation (WIPO) already in 2004, around the beginning of the new century, and therefore at times when private sector started joining the activities in outer space and begun re-establishing the space industry,[11] the need for protection of IPR in outer space started to rise.[12] According to Elon Musk, one of those sharing the vision of sending humans to the Red Planet, the future settlements on Mars should be "self-sustaining",[13] and since they would need all the local resources by themselves, intellectual property might be the only commodity the Martian inhabitants could trade with the inhabitants of Earth.[14] Also, since intellectual ideas regarding artificial intelligence can be considered as intellectual property in common sense generally when these were invented by humans as a legal subject, the topic of intellectual property rights protection in outer space therefore gets more serious only when humans get involved, which can either be a human settlement on Mars or any other celestial body.[15]

In the most countries on Earth, under certain circumstances, innovative ideas with technological background would in most cases be protected as patents which together with ideas connected to design, utility models or arts, geographical indications and trademarks, are commonly known as intellectual property.[16] Despite the fact that each jurisdiction operates under different IPR framework, one remains common to them - the principle of territoriality which shrinks the protection of the intellectual property right for certain territory only.[17] Therefore, the owner of the intellectual property needs to designate each territory on which the IPR protection is desired within a single application under international IPR agreements.[18] Even though humans would

---

[10] *IP WIPO*, 4, mn 18.

[11] Neel V. Patel, '*Are We in the Middle of a New Space Race for This Century?*' (*MIT Technology Review*, 7 February 2020). www.technologyreview.com/2020/02/07/349033/are-we-in-the-middle-of-a-new-space-race-for-this-century/. Accessed 13 September 2020.

[12] *IP WIPO*, 4, mn 18.

[13] Reed Tucker, '*Jeff Bezos and Elon Musk's Plans to Colonize Space Are Even Crazier Than We Thought*' (*New York Post*, 08 August 2020). https://nypost.com/2020/08/08/billionaires-who-plan-to-colonize-space-live-in-a-dream-world/. Accessed 09 September 2020.

[14] As described in: Jackie Wattles, '*Colonizing Mars Could Be Dangerous and Ridiculously Expensive. Elon Musk Wants to Do It Anyway*' (*CNN Business*, 08 September 2020). https://edition.cnn.com/2020/09/08/tech/spacex-mars-profit-scn/index.html. Accessed 09 September 2020.

[15] According to the United States Patent and Trademark Office (USPTO), inventorship is "limited to natural persons" under Title 35 of the United States Code, see USPTO, '*Petition Decision – Inventorship Limited to Natural Persons*'. www.uspto.gov/sites/default/files/documents/16524350_22apr2020.pdf. Accessed 15 September 2020; for the opinion of German Patent and Trade Mark Office (DPMA) see DPMA, '*The Future and the Artificial Inventor*' (3 September 2020). www.dpma.de/english/our_office/publications/background/ai/kuenstlicheintelligenzundschutzrechte/index.html. Accessed 15 September 2020.

[16] WIPO, '*What Is Intellectual Property?*'. www.wipo.int/edocs/pubdocs/en/intproperty/450/wipo_pub_450.pdf. Accessed 15 September 2020.

[17] *IP WIPO*, 4, mn 40; 22, mn 81 f.

[18] WIPO, '*PCT Applicant's Guide*' (1 July 2020). www.wipo.int/export/sites/www/pct/guide/en/gdvol1/pdf/gdvol1.pdf. Accessed 15 September 2020, mn 3.001, 3.005.

not be able to enter into outer space without technology and innovative ideas, outer space is not acknowledged as a territory in the way the territories on the planet Earth are, and therefore, public international agreements can not apply in the same way as they do on Earth. However, it should not be jumped into the conclusion that the boarder of intellectual property rights protection is being marked by the low orbit.[19] Fortunately, a solution for these issues and IPR protection on the Red Planet in general can be addressed in a future Mars Agreement.

2.2 Public International Law and Intellectual Property Law in Context of Outer Space Activities

2.2.1 Sources of Space Law

In order to identify the most effective solution for IPR protection on Mars within a future Mars Agreement, it is necessary to introduce and analyse the already existing legal sources on Space Law and Intellectual Property Law. From the legal perspective, the first question to be asked is which legal framework actually applies in outer space, and whether this includes any regulations on IPR. Despite minor deviations from the common understanding of International Law,[20] Space Law is being considered as a part, or more precisely as lex specialis, to Public International Law.[21] Considering the special status of Space Law within International Law, if any of the Space Law Treaties included regulatory measures on IPR and their protection in outer space activities, these would apply primarily before the international agreements on IPR. Nonetheless, no important conclusions regarding IPR in outer space activities can be made on the basis of the above-mentioned five most relevant agreements which shape the current Space Law. The only source of Space Law which explicitly refers to IPR is the Paragraph 2 of the Declaration by the United Nations Committee on the Peaceful Uses of Outer Space (UN COPUOS) adopted in 1996, which admittedly "introduces new elements into the *corpus juris spatialis*",[22] however actually only

[19] Loren Grush, '*Why Defining the Boundary of Space May Be Crucial for the Future of Spaceflight*' (*The Verge*, 13 December 2018). www.theverge.com/2018/12/13/18130973/space-karman-line-definition-boundary-atmosphere-astronauts. Accessed 15 September 2020.

[20] The main differences are that the responsibility, meaning the extension of jurisdiction on the registered object and the personnel carried by it (*IP WIPO*, 7–8, mn 32), and the liability for objects launched, or even just registered but yet not launched, are conveyed from private enterprises to the launching state and to the state of registry in case that the object causes damage, see Stephan Hobe, *Space Law* (C. H. Beck/Nomos/Hart, 2019), 54–55.

[21] *Id*, 53–55.

[22] Stephan Hobe, Fabio Tronchetti, '*Paragraph 2 (Free Determination of Cooperation)*' in Stephan Hobe, Bernhard Schmidt-Tedd and Kai-Uwe Schrogl (eds), *Cologne Commentary on Space Law* (Volume II, Carl Heymanns Verlag 2015), 335–336, mn 78.

emphasises "the implementation of international cooperation" and underlines the significance of IPR in space technologies.[23]

2.2.2 *Sources of IP Law*

Vice versa, the international agreements which concern international protection of IPR, do not particularly address the protection of IPR in outer space.[24] This legal framework, however, specifies that despite the achieved level of harmonisation of distinct IPR legal regimes within international agreements, the "protection of intellectual property is subject to the applicable territorial legal framework" which constitutes the national and regional IPR legal frameworks which therefore shall be the main sources of IP Law in outer space activities.[25] This corresponds with the most characteristic feature common to intellectual property rights[26]—the principle of territoriality which provides that the holder of the intellectual property right can prevent anybody on the same territory to use and exploit it.[27]

This means two important things in context of outer space activities. Firstly, since the IPR protection is not included in any of the five treaties which constitute the framework for outer space activities, there is no legal framework for IPR protection established through these. Secondly, for the reason that IPR protection is provided by national or regional legal frameworks for the certain territory only and outer space, or even Moon or Mars, do not have any similar legal framework, it seems that an owner of IPR can not protect it on these territories, or more precisely in outer space in its entirety.

Depending on the interpretation, the above-mentioned extraordinary character of Space Law based on the Article VIII of the Outer Space Treaty (OST) can provide solution for IPR protection in outer space together with the Registration Convention[28] through the acknowledgement of the possibility to extend the national or regional laws on IPR of the state of registry to the launching object and, practically, overtaking of the legal responsibility over the launching space object by the state of registry.[29]

---

[23] *Id*, 335, mn 77; *IP WIPO*, 8, mn 34.

[24] *Id*, 5, mn 22 ff.

[25] *Id*, 10, mn 40.

[26] Except for copyright which according to the Berne Convention does not determine the applicable legal framework based on the principle of territoriality but based on the nationality of the author or the authors' residence, see *id*, 12, mn 46. It can be assumed that the question of IPR regimes applicability in outer space therefore would merely become relevant either because of the residency of the author on Mars and elsewhere in outer space or once (if ever) the Martian nationality would be established.

[27] On the most relevant example on patents see WIPO, '*Intellectual Property Handbook*' (WIPO, 2004). www.wipo.int/edocs/pubdocs/en/intproperty/489/wipo_pub_489.pdf. Accessed 14 September 2020, mn 2.3.

[28] *IP WIPO*, 7–8, mn 30–33.

[29] *Id*, 10, mn 41–42.

In this regard, the United States of America and the European Space Agency (ESA) established or acknowledged already existing certain legal frameworks on IPR to be extended to outer space activities as well.[30] Such legal symbiosis of Space Law and IP Law can manifest itself in two different ways, depending on whether the IP originates on a territory on Earth and is launched and used in outer space or whether the IP originates in outer space and is used either there or back on Earth.[31] In this regard, the International Space Station (ISS) represents one of the examples of international cooperation in manned space missions with respect to IPR originating in space.[32] On the other hand, the establishment of the ISS also contributed to the rising questions on the intellectual property rights protection in space and space related activities, as the principle of territoriality and the, although harmonized, but mostly uniquely based national regulations showed the discrepancies between the existing laws on IPR protection.[33] However, even if the Registration Convention as well as OST were interpreted in the way that the local jurisdiction on IPR could apply on a space object, the question remains whether human settlements on Mars should be considered space objects as well. Such problem would not occur, if Mars was either being considered as a new territory, or at least worked under any kind of an international regime.[34] The territory, however, is a problem to which the Mars Agreement also could be the answer.

2.3 Possible Future Scenarios of Intellectual Property Rights Protection on Mars

Even after all the time that has passed since the establishment of the very first international IPR framework in 1883,[35] the opinions on the need for their existence and the regulatory framework still differ. For these reasons, there are basically three options how IPR regulatory frameworks can evolve with regard to future space activities and to the ambitious plans regarding human settlements on Mars in particular.[36] A Mars Agreement would provide the chance for the international community, especially the United Nations Office for Outer Space Affairs (UNOOSA) and

---

[30] *Id*, 11, mn 43; 12, mn 47.

[31] *Id*, 10–11, mn 42.

[32] A.-M. Balsano, '*Intellectual Property Rights and Space Activities*' (*ESA Bulletin*, Nr. 79, August 1994). www.esa.int/esapub/bulletin/bullet79/balsano.htm. Accessed 15 September 2020.

[33] *IP WIPO*, 4–5, mn 19.

[34] See also *id*, 13, mn 52.

[35] WIPO, '*WIPO – A Brief History*'. www.wipo.int/about-wipo/en/history.html. Accessed 08 September 2020.

[36] These can be exemplified by the plans as described in the book Mars Direct by Robert Zubrin and "elaborated in detail" in the book The Case for Mars by Robert Zubrin and Richard Wagner, see Robert Zubrin, '*Elon Musk's Plan to Settle Mars*' (*The Mars Society*, based on the *National Review*, 22 February 2020). www.marssociety.org/news/2020/02/22/elon-musks-plan-to-settle-mars-r-zubrin/. Accessed 09 September 2020; see also Jackie Wattles, '*Colonizing Mars*

UN COPUOS,[37] to determine in which way the future IPR protection in outer space activities should be heading.

2.3.1 Declaration of a Regulation-Free-Zone as a Contribution to Innovation

If one would follow the opinion that intellectual property is not fostering innovation,[38] the logical consequence would be to declare Mars as an IPR regulation-free-zone within a future Mars Agreement, meaning that there would be no particular IPR regime established on this territory. In this case, either the status-quo would remain and the IPR regulatory frameworks together with the national jurisdictions would apply further in the same way as described above, or alternatively, the non-applicability of any IPR protection on Mars could be declared.

Nevertheless, since more private companies are participating in outer space activities, the space race is not characterised by the scientific and security aspect as much as it used to. Since the private investment and the own know-how need to be secured from competitors, the option of a regulation-free zone on Mars and on other celestial bodies therefore does not seem to be realistic. Moreover, it would probably move some of the players from the commercial sector to either completely discontinue participating or reduce their activities in the outer space race and also prevent new ones from even joining. An argument in this regard could be made that even some private companies do not file for a patent and therefore do not seek for patent protection on their inventions anymore but protect their innovative ideas as trade secrets.[39] The problem however is merely that in order for an invention to be protected under patent laws, it needs to be made public so that the scientific community knows the state of the art.[40] The interest in IPR protection therefore remains.

Could Be Dangerous and Ridiculously Expensive. Elon Musk Wants to Do It Anyway' (*CNN Business*, 08 September 2020). https://edition.cnn.com/2020/09/08/tech/spacex-mars-profit-scn/index.html. Accessed 09 September 2020.

[37] UNOOSA, '*Roles and Responsibilities*'. www.unoosa.org/oosa/en/aboutus/roles-responsibilities.html. Accessed 15 September 2020.

[38] An example for pro and contra considerations around patent protection in *The Great IP Debate: Do Patents Do More Harm Than Good?*' (*Science|Business*, 28 July 2016). https://sciencebusiness.net/news/79887/The-Great-IP-Debate%3A-Do-patents-do-more-harm-than-good%3F. Accessed 10. September 2020; "unprecedented criticism" of the patent protection system in the USA in particular described in Maureen K. Ohlhausen, '*Patent Rights in a Climate of Intellectual Property Rights Skepticism*' (*Harvard Journal of Law & Technology*, Volume 30, Number 1, Fall 2016). https://jolt.law.harvard.edu/assets/articlePDFs/v30/30HarvJLTech103.pdf. Accessed 10 September 2020, 110 ff.

[39] Kim Bhasin, '*Elon Musk: If We Published Patents, It Would Be Farcical*' (*Business Insider*, 9 November 2012). www.businessinsider.com/elon-musk-patents-2012-11?r=DE&IR=T. Accessed 15 September 2020; WIPO, '*Intellectual Property Handbook*' (WIPO, 2004). www.wipo.int/edocs/pubdocs/en/intproperty/489/wipo_pub_489.pdf. Accessed 15 September 2020, mn 2.837.

[40] *Id*, mn 2.5.

2.3.2 *Establishment of a New Public International Law Agreement Regarding Intellectual Property Rights in Space Activities*

The second option which could be included in a future Mars Agreement regarding IPR protection in outer space would be the establishment of an international agreement similar to those administered by WIPO.[41] Since these treaties and agreements provide effective mechanisms and systems for global protection of IPR on Earth, it seems natural to establish a new public international treaty for outer space activities as well. The most convincing aspect about such agreement would be that it might provide a certain IPR regime not only for Mars but for the whole outer space. A Mars Agreement could therefore explicitly state that such IPR agreement shall apply at the Martian territory. As such, it could either establish a regime similar to the Patent Cooperation Treaty (PCT), the Madrid Agreement and others, or a unified system, similar to the European Trademark and Design, including the latest attempt for the European Unified Patent. Since the concepts show considerable differences, it should be looked at these more closely.

Within the international IPR regime established by the international public agreements and treaties, the most important aspect in this regard is the particular role of WIPO. In principle, WIPO serves as a central point of contact for all parties which seek cross-boarder IPR protection under the treaties administered by WIPO.[42] As mentioned above, the treaties establish such a regime and mechanism that the applicant only needs to file one application and to point out in which countries such protection is wished to be granted. With regard to the principle of territoriality and the national IP protection regulations, the protection only is granted if the national office decides so. Therefore, the roles within the procedure of granting IPR protection differ, the role of WIPO being a more formal one, and the role of the national offices being of a more material nature. However, this means that without WIPO and the treaties, the applicant would need to file for IP protection individually on every territory, i. e. usually in every country, where such protection is sought.

If a similar treaty applied for Mars (or the Moon or other celestial body), the role of WIPO would for the time being remain the same. It would be more likely that the inhabitants of Mars would not participate in the jurisdiction existing on Earth in the same way. Moreover, they might establish their own jurisdiction and rules, in this particular case even their own IP Law. If this happened, the inhabitants of Mars could establish their own national office for such purposes, which could be participating in the system already existing on Earth, also in an agreement considering IPR protection on Mars in particular. WIPO could still act within the formal part of

---

[41] For the list of treaties providing IP protection and Global Protection System administered by WIPO see WIPO, '*WIPO-Administered Treaties*'. www.wipo.int/treaties/en/ . Accessed 07 August 2020.

[42] To the role of WIPO and Intellectual Property in general see WIPO, '*Intellectual Property Handbook*' (WIPO, 2004). www.wipo.int/edocs/pubdocs/en/intproperty/489/wipo_pub_489.pdf. Accessed 15 September 2020.

the proceedings, as the agent between the national office of a territory on Earth and the territory on Mars. According to the above-said, the protection would, however, only be granted if a national (i. e. probably Martian) office for IP protection decided so in accordance with local law.

The other direction for a new public international agreement might be the establishment of an unified IPR regime. For this reason, such ideas as the establishment of an "Intellectual Property Galaxy Treaty"[43] or "Space Patent"[44] have started to rise recently. In theory, a unified framework on IPR provides more efficiency and therefore promises benefits such as time and cost saving.[45] However, it might be a tedious process to get there from a practical point of view. An example for the implementation of the idea of the establishment of a whole new unified framework is the attempt to unify the IPR framework within the European Union (EU). Despite the fact that the system regarding community trade mark and design has been working successfully for 25 years,[46] a similar initiative towards a unified patent system within the EU still faces obstacles.[47]

On the other hand, even though some actors of the private sector have very ambitious plans regarding human settlement on Mars in the upcoming years, it is possible that humans settle on the Moon or other celestial body earlier than on Mars.[48] Or to be precise, there is a possibility that humans settle on more celestial bodies than Mars in the future which means that the necessity of IPR protection in space activities would rise repeatedly. Such agreement could therefore include not only Mars as a new territory but other celestial bodies, objects in the orbit around the Earth or the orbit around the Moon or other celestial bodies as well. This would be an asset considering the trend of growing numbers of participants in space activities, especially those from the private sector. If the scope of such agreement was that comprehensive, the demanding work probably still would be worth it as it would provide a long-term solution to more territories in outer space than just Mars itself.

---

[43] A. M. Davidson, '*To Explore Outer Space: The Intellectual Property Frontier for Patents*' (*Hofstra Law Review*, Vol. 47:889, 2019). www.hofstralawreview.org/wp-content/uploads/2019/05/bb.3.davidson.pdf. Accessed 06 August 2020.

[44] Yun Zhao, '*Protection of Intellectual Property Rights in Outer Space*' (*American Institute of Aeronautics and Astronautics, Inc.*, 2006). https://iislweb.org/docs/Diederiks2006.pdf. Accessed 13 September 2020.

[45] Based on the example of the Unitary Patent, see further European Patent Office, '*Unitary Patent*'. www.epo.org/law-practice/unitary/unitary-patent.html. Accessed 15 September 2020.

[46] Beginning in 1994 as Community Trade Mark and nowadays continuing as European Union Trade Mark, see further European Union Intellectual Property Office, '*Our History*'. https://euipo.europa.eu/ohimportal/cs/our-history . Accessed 11 September 2020.

[47] See further '*Press Release No. 20/2020*' (20 March 2020). www.bundesverfassungsgericht.de/SharedDocs/Pressemitteilungen/EN/2020/bvg20-020.html. Accessed 11 September 2020; more in detail '*Order 2 BvR 739/17*' of the Federal Constitutional Court of Germany. www.bundesverfassungsgericht.de/SharedDocs/Entscheidungen/DE/2020/02/rs20200213_2bvr073917.html; jsessionid=ADBE1702990A7CC810C12D169D71E9D2.2_cid394. Accessed 11 September 2020.

[48] Jan Woerner, '*A Vision for Global Cooperation and Space 4.0*'. www.esa.int/About_Us/Ministerial_Council_2016/Moon_Village. Accessed 10 September 2020; further information to The Moon Village at https://moonvillageassociation.org.

2.3.3 *Acknowledgement as a New Territory and Extension of Currently Existing Intellectual Property Rights Agreements*

The third option which could possibly be included in a Mars Agreement is the acknowledgement of Mars as a new territory on which the already existing international IPR agreements and treaties apply. For this reason, the Red Planet would basically need to be declared as a territory in the sense of these already existing agreements. If humans, at least after certain time after they have settled on Mars, established their own legal regime, they still could become parties to the already existing international IP agreements as well.

In a similar way as stated above, the extension of these treaties to the territory of Mars would mean that it would then be necessary to establish an institution such as a "national" office for IPR matters or to cooperate with an already existing one which would actually take over the IPR agenda for the Martian territory. For this, the acknowledgement of an existing IPR protection regime would be needed as well. The IPR protection would then apply in the same manner as on territories of Earth which are parties to the IP international treaties and agreements administered by WIPO.

2.4 Conclusion

Since neither the Moon Agreement nor any other existing agreement on outer space related activities include the topic of IPR protection on celestial bodies or in outer space activities, a Mars Agreement could become the first international agreement which includes regulations on IPR on a celestial body in outer space. As discussed above, the tendency to abandon any IPR protection in outer space activities seems to be little. Therefore, the establishment of a new public international IPR treaty or the extension of the already existing ones on the declared territory of Mars could be included as regulations on the IPR protection on Mars within a future Mars Agreement. For these purposes, it would not be necessary to define an entirely new regulatory framework for Mars. Certainly, it would be sufficient if the trend and the direction of the future IPR protection were declared. Even though only the parties to a Mars Agreement would be bound by it, such clause could also serve as an example and a basis for any future IPR agreement.

In conclusion, a future Mars Agreement could therefore address and provide direction in the matter of IPR protection on Mars, established by the international space community and based on its experience and future plans in outer space activities. With regard to the very ambitious plans concerning future activities in outer space of both, the public as well as the commercial sector, taking a step in one of the above-mentioned two directions, or maybe even pursuing a combination of both of them, seems to be inevitable in the future. A clear statement within a Mars Agreement towards one of these solutions might provide the already participating actors but also the newcomers to the space sector with more legal certainty and encourage them to further engage in their ambitious plans around human settlements on Mars. Despite the fact that the scope of these solutions would differ, a solid solution for IPR protection on the Red Planet (possibly and eventually even beyond) would be provided.

Terezie Němcová is a law student at the University of Erlangen-Nuremberg in Germany. In addition to her university studies of German Law, she seeks knowledge and gains practical experience in the field of Space Law and Intellectual Property Law, with particular interest in the legal aspects of intellectual property rights protection in outer space activities.

Chapter 3
On the Province of All Marskind

Maria Lucas-Rhimbassen

He who controls spice controls the Universe.
Frank Herbert, Dune.

Abstract The purpose of this article is to explore the challenging context surrounding the non-appropriation principle as recent developments in space law take an unexpected turn towards accelerated commercialization and blurring ethical principles. By analyzing the reinterpretation of the "*res communis*" regime of outer space, the reader understands that this regime is in transition and gradually slides towards private property rights and trends despite the non-appropriation principle. This raises questions as to competition law rights between private stakeholders (firstcomers *vs* newcomers) and potential monopolies restricting access to critical resources and services. Such commercial activity must further be closely monitored since it might involve modifying the core identity of a celestial body *per se*, as in the case of terraforming. Space ethics could provide guidance against such scenarios until an appropriate regulatory regime is agreed upon at the international level. Last, proper antitrust regulation is vital for a prosperous Mars settlement.

3.1 Introduction

In the light of the *corpus juris spatialis*, it was considered rather unanimously until recently that space was a realm falling under the "*res communis*" regime. Some scholars refute this while interpreting restrictively the Outer Space Treaty (OST) Article I which posits that outer space is a "province of all mankind". These scholars qualify outer space *per se* as accessible to all indeed, but not as *res communis*. Rather, this would apply to the commercial space activities conducted by spacefaring nations and, as of this writing, increasingly more by private actors. Furthermore, the *res communis* regime based on activities which should be conducted in the interest of all mankind and instill benefit-sharing principles as subsequently elaborated in the Space

M. Lucas-Rhimbassen (✉)
Chaire SIRIUS, IDETCOM, University of Toulouse, Toulouse, France
e-mail: maria.lucas-rhimbassen@ut-capitole.fr

A. Froehlich (ed.), *Assessing a Mars Agreement Including Human Settlements*, Studies in Space Policy 30, https://doi.org/10.1007/978-3-030-65013-1_3

Benefits Declaration of 1996,[1] is once more debated. Indeed, several scholars diverge. On one hand, the Hague International Space Resources Governance Working Group concluded in its Building Blocks[2] that space is accessible to all, equally, but benefits should not include financial benefit-sharing derived from space activities since that would deter private actors from undertaking colossal endeavors characterised by lower return on investment (ROI). On the other hand, the Outer Space Insitute's Vancouver Recommendations[3] voiced another opinion, which, on the contrary would include such financial sharing to benefit all nations, equally. The former position reflects a more holistic group, comprised of academia, public institutions and private actors such as industry and start-ups. The latter position reflects a more scholarly nature. This debate will likely continue at the intellectual, ethical and legal level, possibly at international venues such as the UNCOPUOS, however, facts seems to embrace a market-centric approach.

3.2 The Space Tuna Paradox

In their *Treatise*,[4] Lyall and Larsen wrote that fish in high seas cannot be owned yet the profit of their sale can be:

> Title is recognised to fish, ownerless when in the high seas, but readily sold by fishermen on the shore.

This translates into the nascent space law interpretation of space resources appropriation, although, paradoxically, fish in high seas qualify as *res communis*. With respect to the non-appropriation principle enshrined in Article II of the OST, this is significant since several national space legislations, such as the US Space Act of 2015,[5] the Luxembourg *Act on the Exploration and Use of Space Resources* (the

---

[1] 51/122. Declaration on International Cooperation in the Exploration and Use of Outer Space for the Benefit and in the Interest of All States, Taking into Particular Account the Needs of Developing Countries, UNOOSA, https://www.unoosa.org/oosa/en/ourwork/spacelaw/principles/space-benefits-declaration.html. (Retrieved on 2nd October, 2020).

[2] Building Blocks for the Development of an International Framework on Space Resources Activities, November 2019, https://www.universiteitleiden.nl/binaries/content/assets/rechtsgeleerdheid/instituut-voor-publiekrecht/lucht--en-ruimterecht/space-resources/bb-thissrwg--cover.pdf. (Retrieved on 4th October, 2020).

[3] The Vancouver Recommendations on Space Mining, OSI, http://www.outerspaceinstitute.ca/docs/Vancouver_Recommendations_on_Space_Mining.pdf.

[4] Lyall and Larsen, 'Space Law: A Treatise', Routledge, 2017.

[5] The *US Space Act of 2015* or *Title IV of the Commercial Space Launch Competitiveness Act of 2015 (CSLCA)* states that: "(Sec. 402) The bill directs the President, acting through appropriate federal agencies, to:

- facilitate the commercial exploration for and commercial recovery of space resources by U.S. citizens;

Space Resources Act) of July 2017[6] and the United Arab Emirates (UAE) national space law including provisions pertaining, among others, to space resources (Article 18 on *Exploration, Exploitation and Use of Space Resource)*[7] all give right to their nationals to commercially exploit space resources while formally denying unilaterally the global commons regime. For instance, this is illustrated by the 2020 US *Executive Order on Encouraging International Support for the Recovery and Use of Space Resources* (EO)[8] which sets the tone for the nation's ambitions to further unilateral efforts with regards to spacefaring activities. Section 1 provides *inter alia* that:

> Americans should have the right to engage in commercial exploration, recovery, and use of resources in outer space, consistent with applicable law. Outer space is a legally and physically unique domain of human activity, and the United States does not view it as a "global commons". Accordingly, it shall be the policy of the United States to encourage international support for the public and private recovery and use of resources in outer space, consistent with applicable law.

Moreover, the EO rejects the "Global Commons" *status quo* regime and "common heritage of mankind" as set in the 1979 Moon Agreement (MA).[9] formally rejected by the US through the EO. This is particularly relevant in the context of the above-mentioned section since the Moon Agreement further elaborates on space resources, and more importantly, on the prevention of harmful contamination of celestial bodies at a more holistic level. If not rejected, the MA could have set a foundational precedent which would be relevant in the case of Mars, as a starting level, especially with regards to terraforming projects in the future.

- discourage government barriers to the development of economically viable, safe, and stable industries for the commercial exploration for and commercial recovery of space resources in manners consistent with U.S. international obligations; and
- promote the right of U.S. citizens to engage in commercial exploration for and commercial recovery of space resources free from harmful interference, in accordance with such obligations and subject to authorization and continuing supervision by the federal government.
- a U.S. citizen engaged in commercial recovery of an asteroid resource or a space resource shall be entitled to any asteroid resource or space resource obtained, including to possess, own, transport, use, and sell it according to applicable law, including U.S. international obligations"; etc. (SPACE ACT OF 2015).

[6]Which states that "Space resources are capable of being owned".

[7]Which states that: "Subject to the provisions of Article (14) of this Law, the conditions and controls relating to Permits for the exploration, exploitation and use of Space Resources, including their acquisition, purchase, sale, trade, transportation, storage and any Space Activities aimed at providing logistical services in this regard shall be determined by a decision issued by the Council of Ministers or whomever it delegates".

[8]Executive Order on Encouraging International Support for the Recovery and Use of Space Resources, Executive Order, 6 April 2020, https://www.whitehouse.gov/presidential-actions/executive-order-encouraging-international-support-recovery-use-space-resources/. (Retrieved on 30th September, 2020).

[9]Agreement Governing the Activities of States on the Moon and Other Celestial Bodies, UNOOSA, 1979, https://www.unoosa.org/oosa/en/ourwork/spacelaw/treaties/intromoon-agreement.html. (Retrieved on 30th September, 2020).

Table 3.1 NASA mission planetary protection

Lunar target sensitivity	Mission PP category
Not of direct interest for understanding the process of chemical evolution or where exploration will not be jeopardized by terrestrial contamination. No protection of such regions is warranted	I-L
Of significant interest relative to the process of chemical evolution but only a remote chance that biological contamination by spacecraft could compromise future investigations. Reporting of biological materials is warranted	II-L

Nonetheless, even though the MA is being discarded, harmful contamination is still addressed internationally under the OST, Article IX, and unilateral efforts are being made at the national level to define measures against harmful contamination by private actors while conducting commercial activities on celestial bodies such as the Moon or Mars. For example, NASA's Interim Directives (NIDs)[10] divide the lunar surface in two main categories to mitigate forward contamination at best. This is illustrated in the following Table 3.1.

This is a prelude to other timely directives awaiting in the case of Mars since there is potential for bio-signature patterns and planetary protection measures. Guidelines need to be taken seriously especially since researchers have discovered phosphine in the Venus atmosphere and plan to send probes to take samples.[11] Therefore, both forward and backward harmful contamination procedures are becoming a priority despite the non-binding Committee on Space Research (COSPAR) Planetary Protection Policy indications.[12] This is relevant to the present paper since events in the space sector are accelerating. On one hand, within the timespan of one week, researchers discovered potential signs of life on neighbouring celestial bodies, while, NASA, one the other hand, officially allowed the commerce of abiotic space resources on the Moon. More precisely, NASA is to incentivize commercial extraction of lunar regolith, water ice, precious metals, rare earths, etc. and is to purchase symbolic sums at symbolic prices from commercial actors to initiate the space resources economy and resulting space exchange market.[13] With the NASA Artemis Program and the Commercial Lunar Payload Services (CLPS), it is likely that this will trigger a competitive ecosystem among New Space actors. The backbone of this article

[10]NASA Interim Directive (NID) Effective Date: 9 July, 2020 Expiration Date: 9 July, 2021 Planetary Protection Categorization for Robotic and Crewed Missions to the Earth's Moon NID 8715.128 https://nodis3.gsfc.nasa.gov/OPD_docs/NID_8715_128_.pdf at 5. (Retrived on 30th September, 2020).

[11]Jonathan Amos, Venus: Will private firms win the race to the fiery planet?, BBC.com, https://www.bbc.com/news/science-environment-54151861, 14 September 2020.

[12]Thomas J Herron, 'Deep Space Thinking: What Elon Musk's Idea to Nuke Mars Teaches Us about Regulating the Visionaries and Daredevils of Outer Space' (2016) 41 Colum J Envtl L 553.

[13]Jeff Foust, 'NASA offers to buy lunar samples to set space resources precedent', Space News, 10 September 2020, https://spacenews.com/nasa-offers-to-buy-lunar-samples-to-set-space-resources-precedent/. (Retrieved on 20th September, 2020).

addresses, in the next section, the need for an authority to regulate such competition on celestial bodies to (1) avoid future monopolies and; (2) to monitor natural monopolies when they are the only option.

Last, when interviewed in 2020, NASA Administrator, Jim Bridenstine compared outer space resources to "tuna", a comparison which is reminiscent of Lyall and Larsen's mention of the fish in high seas. Bridenstine stated:

> But we do believe that we can extract and utilize the resources from the moon, just like we can extract and utilize tuna from the ocean (...) "And so the question is, Is it possible to have property rights for extracted resources without appropriating the moon or other celestial bodies for national sovereignty? And I believe that the answer is overwhelmingly yes".[14]

While rejecting the commons doctrine, he nevertheless relied on it which is contradictory in legal terms. However, Lyall and Larsen admitted in their *Treatise* that, ultimately, the nature of common goods in outer space will be subject to change in time:

> Position as to materials from the Moon and celestial bodies may, however, be different or may be different in the future.

Apparently, this time has come. Consequently, legal innovation can be under way, but universal consensus is not in sight and therefore, priority is shifted towards soft law and principles, such as in the case of the Artemis Accords' set of principles. Hence, there is a growing attention granted to space ethics and their guiding role in these changing times, especially to ensure equal access to space and opportunity to compete, and future activities such as terraforming, which is the ultimate goal of a Martian "settlement".

3.3 Martian Market and Monopolies

The previous section determined the challenging position of space law with regards to the non-appropriation principle which appears to go together with a nascent trend of space property doctrine such as applying the Anglo-American principle of "finders-keeper" on celestial bodies in terms of "first in time, first in right" or "first possession" rights to exploit space resources.[15] To encourage the growth of such industry, private actors could claim to be entitled to the benefits resulting from resource extraction to justify the astronomical costs they would have to invest in a market with very high entry barriers. On top of that, the first wave of actors or "firstcomers" could seek to maximize their profit and secure an important market share close to a monopolistic status. If there are no other competitors on a given field and if the economies of scale and efficiency are justified, after having heavily invested in infrastructure, one actor

[14] Ibid.

[15] David Collins, 'Efficient Allocation of Real Property Rights on the Planet Mars' (2008) 14 BU J Sci & Tech L 201.

can attain the position of natural monopoly.[16] In that case, it could either contribute to the expansion of the specific market by selling the products, services or access to new coming actors or, it could seek to preclude them from that same market by harvesting more resources and commodities than required, increasing in this manner prices, anticompetitive behavior and discriminatory practices. These dynamics are characteristic to the commercialization of services. Increasing privatization of space may enable such scenarios and endanger the benefit-sharing principles enshrined within the OST. In that case, it is vital to implement a framework based on ethics, competition law and antitrust with regards to resources utilization, market and settlement on celestial bodies such as Mars. More importantly so when these commercial services involve changing the environmental pattern of an entire planet, such as in the case of terraforming which require further holistic consensus and eco-systemic cohesiveness.

In the case of Mars, early commercial actors or "firstcomers" could, according to certain scholars, "gain control of access" to Mars and if unchecked, they could be tempted to control the access of newcomers to critical resources. Their size and power could even procure them with unmatched self-sustainability in that context, similar to the "*Transnats*" depicted in the sci-fi Mars trilogy by Kim Stanley Robinson. For instance, on Mars, water and carbon dioxide can be converted into rocket propellant or used as water. Therefore, accessing water or other commons or commodities—depending on their future legal status—can be subject to commercialization and therefore to "unnatural" monopolies/oligopolies if no authority regulates the market:

> "If the firstcomer develops resource gathering on Mars and no second comers establish a competitive position, a single entity could effectively exclude others, including sovereign nations or coalitions thereof, from accessing Mars by cutting off their supply of these vital resources on which their mission architectures will likely rely (...) If a mission to Mars relies on the purchase of water, carbon dioxide, oxygen, or rocket propellant from the monopolist and the monopolist refuses to meet the mission's needs, the group would have no alternative supply sources. In that case, the group might not be able to make it back to Earth safely. Additionally, in a worst-case scenario, the firstcomer's Martian settlement could begin to exist as a self-sustaining society outside the reach of Earth-based regulation and influence. Under the current international model of national non-intervention in space, there would be no basis for a sovereign government to enforce any type of legal judgment".[17]

Future human settlements on Mars would require natural monopolies to be properly regulated to secure a competitive but sustainable market in due time. Moreover, when private actors look into terraforming (e.g. SpaceX), it is vital to preclude them from creating terraforming monopolies since *them who control terraforming on a celestial body, control its very identity* and may ultimately encounter the risk of consequential hubris. This may indeed be detrimental from a policy perspective.

[16]Richard Posner, 'Natural Monopoly and its Regulation', Cato Institute, 1999.

[17]Tyler Conte, 'Property Rules for Martian Resources: How the Space Act of 2015 Increases the Likelihood of a Single Entity Controlling Access to Mars' (2019) 84 J Air L & Com 187 (CONTE).

3.4 Terraforming: Consensual Ethics or Competing Hubris

Article 7.1 of the previously mentioned Moon Agreement explicitly forbids modifying the environment of a celestial body up to the point of altering its systemic identity:

> In exploring and using the Moon, States Parties shall take measures to prevent the disruption of the existing balance of its environment, whether by introducing adverse changes in that environment, by its harmful contamination through the introduction of extra-environmental matter or otherwise. States Parties shall also take measures to avoid harmfully affecting the environment of the Earth through the introduction of extraterrestrial matter or otherwise.

However, the MA is being gradually discarded and no such clause remains in international space law with regards to celestial bodies, let alone Mars. Since law remains silent, in addition to the precautionary principle set out in the Rio de Janeiro Convention of 1992[18] and its Articles on Transboundary Harm[19] and the Liability Convention of 1972,[20] one can turn to space ethics for further guidance and in particular to terraforming ethics which are the following[21]:

> Any system of environmental ethics is based on three fundamental axioms: anti-humanism, wise stewardship, and intrinsic worth.

Anti-humanism rejects using technology to alter the environment since it is believed to be too complex. Therefore, in this case, it would be against terraforming Mars and changing the ecology of a planet. On the contrary, "wise stewardship", also known as utilitarianism, holds that "humans can use and alter natural systems but must do so wisely and to the long-term benefit of humanity". This view is the most popular in environmental ethics. Finally, intrinsic worth debates as to what is more important: animate or inanimate, Terrestrial or Martian? It is the most subjective view. Further ethical categories include "geocentrism", "ecocentrism", "zoocentrism" and "anthropocentrism" or "homocentrism" but new ethical approaches involve "cosmic preservationism", "restorationism", and "inventionism", which is the most conducive to terraforming besides "anthropomorphism".

---

[18]United Nations Conference on Environment and Development (UNCED), UN, https://www.un.org/esa/dsd/resources/res_docukeyconf_eartsumm.shtml. (Retrived on 2nd October, 2020).

[19]Draft articles on Prevention of Transboundary Harm from Hazardous Activities, with commentaries 2001, UN, 2001, https://legal.un.org/ilc/texts/instruments/english/commentaries/9_7_2001.pdf. (Retrived on 4th October, 2020).

[20]Convention on International Liability for Damage Caused by Space Objects, UNOOSA, https://www.unoosa.org/oosa/en/ourwork/spacelaw/treaties/introliability-convention.html. (Retrieved on 4th October, 2020).

[21]Robert D Pinson, 'Ethical Considerations for Terraforming Mars' (2002) 32 Envtl L RepNews & Analysis 11333.

3.5 Discussion

While growing focus is directed towards private property rights on celestial bodies due to the increasing privatization of the sector and commercialization of services, some authors recommend an international regulatory body to look into appropriate property rights categories which might optimize private space endeavors and a thriving interplanetary economy. For instance, it has been proposed to advocate for a new regime of "Riparian Rights Rule", under which stakeholders should comply with sustainability standards and equality of access to resources and ensure a model of "appropriation that better protects the benefits of Mars for all humankind".[22] This would entail:

> (1) a licensing regime for resource production on Mars that would be granted by a coalition of sovereign governments under an international treaty; (2) price regulation for the sale of resources by the first producer that has established a natural monopoly-like advantage on Mars, which would, to a certain degree, resemble the rate-making process for vertically integrated electricity providers in the United States; and (3) the Martian Riparian Rights Rule, under which existing producers of Martian resources would not have the right to extract more than necessary for their own use or for sale to current or foreseeable future consumers.

Conte further suggests amending the 2015 SPACE Act's[23] property rule to:

> preserve second comers' opportunities to compete with the firstcomer, thereby preventing the first-moving monopolist from effectively controlling access to the planet through its provision of water, oxygen, and rocket propellant to all second comers.

This measure of competition law at the national level could indeed prove to be efficient and constitute a precedent which might gather critical mass and momentum at the international level once the industry is rolling and that the space community needs adapting. This could prevent monopolies, but could generate oligopolies and competition law dynamics which would require transnational or even "trans-planetary" regulation—instead of supranational, which is currently being challenged at the global governance level.

Another proposal suggests that celestial bodies be considered as *terra nullius* to further secure property rights and incentivize settlements. Brandon[24] suggests that celestial bodies be subject to the *res nullius* doctrine rather than *res communis* which would attract private actors to invest, much like in the case of early "colonization" and early "settlers" in the United States of America, pushing the final frontier further away. He coined his proposal as "*res nullius humanitatus*" which may resonate with the commercialization trend, but which might not generate consensus internationally. Space law faces increasing challenges pertaining to property rights in space as the age of space commerce beckons.

---

[22] CONTE, supra note 18.

[23] SPACE ACT OF 2015, supra note 6.

[24] Brandon C Gruner, "A New Hope for International Space Law: Incorporating Nineteenth Century First Possession Principles into the 1967 Space Treaty for the Colonization of Outer Space in the Twenty-First Century" (2004) 35:1 Seton Hall L Rev 299.

3.6 Conclusion

Space law, national space legislation, non-appropriation, space commercialization, New Space, space resources in situ utilization and commerce, property law, competition law, planetary protection and terraforming are all converging towards a nexus of legal void which needs regulation in time. Otherwise, this void might be filled exclusively by lex mercatoria and leave little room for ethics. Adaptive governance must ensure that the foundations of the most appropriate framework are set in motion and begin already to frame customary law and practice which seem rather chaotic as of this writing. Until then, ethics provide an important guiding point thanks to their broader range and shared acceptance through society at large.

Proposals are being made which pinpoint to a hardcore private industry in space, based on either "first time, first in right" or concession allocations which rely private property. This confronts the basis of international space law which precludes ownership by occupation and the fact that, under lex spatialis, no State has the authority to grant title over celestial bodies. However, as with the Artemis Accords, private entities will go up there and proceed, if profitable, to resource extraction and commerce. On Mars, that would sooner or later entail efforts towards terraforming and controlling the red planet, which requires an international framework. If no effort is undertaken as of now, further unilateral initiatives might manifest and challenge fundamental principles of the corpus juris spatialis, such as benefit-sharing and equality. It is already the case with the Artemis Accords. A Martian Accords scenario could indeed be more extreme and therefore a new regime for the benefit of all "Marskind" must be studied.

Maria Lucas-Rhimbassen is a Ph.D. Candidate in Space Law at the Chaire SIRIUS (University of Toulouse, France) where she conducts graduate research on national space legislation, critical space infrastructure resilience, on-orbit insurance and liability, and space antitrust issues. She provides consultation services to public and private stakeholders such as CNES, Airbus and Thales Alenia Space on a variety of topics. Ms Lucas-Rhimbassen holds graduate degrees in management from HEC Montreal (Canada) and in law from the University of Moncton (Canada) and Le Havre (France) and an executive-level certificate in Strategic Space Law from the IASL at McGill University (Canada).

Chapter 4
How Five Fundamental Human Rights Could Be Violated in Privately-Funded Space Settlements and the Role of the Mars Agreement in Their Protection

Juan García Bonilla

Abstract It is becoming increasingly clear that human settlements are part of the future of the space sector. With the advent of NewSpace, it is also clear that as long as there are potential economic benefits, private companies will play a key role in the habitation of space. This article focuses on answering two questions: how five rights recognized in the Universal Declaration of Human Rights could be threatened in privately-funded space settlements and what provisions should the Mars Agreement make to protect them. In particular, the ban on arbitrary detentions, the right to privacy, the right to a nationality, the right to marriage and family, and the right to freedom of assembly and association are discussed.

4.1 Introduction

Science fiction has painted a grim picture for corporations that get involved in the futuristic business of "space colonization". In the horror classic *Alien*, the megacorporation *Weyland-Yutani* has little regard for the life of its settlers in their evil quest to weaponize the xenomorph. In the generation-defining universe of *Star Wars*, the *Corporate Sector Alliance* is a coalition of private corporations that tyrannically rule thousands of planets. Needless to say, slavery and the destruction of natural habitats are common practices in said planets. While science fiction is, after all, still fiction, there might be genuine reasons why approaching the issue of privately-owned space settlements must be done with great care.

For a less fictional precedent, the history of the East India Company may be studied. An initially privately owned corporation, it came to directly control the greater part of the Indian Subcontinent from the seventeenth to the nineteenth century. Under the administration of the company, territories were exploited, and

J. García Bonilla (✉)
Universidad Carlos III de Madrid, Leganés, Spain
e-mail: juan@garciabonilla.com

A. Froehlich (ed.), *Assessing a Mars Agreement Including Human Settlements*, Studies in Space Policy 30, https://doi.org/10.1007/978-3-030-65013-1_4

the wellbeing of the population disregarded while generating great profits for its shareholders.[1] In William Dalrymple words:

> We still talk about Britain conquering India. [...] It was not the British Government that seized India at the end of the 18th Century, but a dangerously unregulated private company, headquartered in one small office, five windows wide...[2]

Even though dystopian science fiction tends to err on the side of catastrophism, history proves that unchecked companies with too much control over a population tend to abuse their positions of power. It is worrisome that private human settlements in space could provide the perfect scenario for new and dangerous forms of colonialism to proliferate. In this article, the role of the Mars Agreement in the protection of human rights in privately-owned settlements is explored.

4.2 The Future of Space: Permanent Human Settlements

Humanity has learned to adapt and occupy every ecological niche on Earth, yet its ambition for expansion has not dwindled. National space agencies around the world have shown interest over the possibility of occupying space, and efforts are underway to explore the Moon and Mars and to discover new ways to make them habitable. Likewise, private companies have turned their eyes towards space, considering the profitability of investing in the space sector.

NASA's Artemis program's long-term goals include the creation of a surface habitat on the moon and the construction of a habitable mobile platform.[3] The Mars Exploration Program, on the other hand, hopes to provide insights into the Red Planet's environment, and demonstrate technology that would make its habitation possible, such as Perseverance's MOXIE instrument, which will transform martian CO_2 into O_2.[4,5]

Other national space agencies have similarly expressed interest in setting up long-term habitats in space: ESA's Moon Village concept envisions a community of scientists and private investors setting up stations closely together to exploit potential

[1] William Dalrymple, 'The East India Company: The original corporate raiders' *The Guardian* (4 March 2015) https://www.theguardian.com/world/2015/mar/04/east-india-company-original-corporate-raiders. Accessed 12 September 2020.

[2] William Dalrymple, 'The East India Company: The original corporate raiders' *The Guardian* (4 March 2015) https://www.theguardian.com/world/2015/mar/04/east-india-company-original-corporate-raiders. Accessed 12 September 2020.

[3] Brian Dunbar, 'Moon to Mars Overview' (*National Aeronautics and Space Administration website,* 10 September 2020) https://www.nasa.gov/topics/moon-to-mars/overview. Accessed 12 September 2020.

[4] National Aeronautics and Space Administration, 'Mars Exploration Program' (*National Aeronautics and Space Administration website*) https://mars.nasa.gov/. Accessed 12 September 2020.

[5] National Aeronautics and Space Administration, 'Mars Oxygen ISRU Experiment' (*National Aeronautics and Space Administration website*) https://mars.nasa.gov/mars2020/spacecraft/instruments/moxie/for-scientists/. Accessed 12 September 2020.

synergies between these groups.[6] The China National Space Administration, on the other hand, has its sights on the commercial utilization space. In 2002, Ouyang Ziyuan, chief scientist of the Chang'e program, stated that "[t]he moon could serve as a new and tremendous supplier of energy and resources for human beings […] Whoever first conquers the moon will benefit first".[7]

Even more relevant for this article's topic, several CEOs of private companies have publicly expressed their interest in founding human settlements in space. Elon Musk's vision for a Mars settlement includes one million humans living in it by 2050, and 100 Starships being built every year for 10 years in order to make the necessary trips from and to the planet.[8] Jeff Bezos' idea for permanent settlements in space, however, involves the creation of O'Neill cylinders: mega-structures floating in space that rotate to simulate terrestrial gravity.[9]

It is impossible to predict who will be the first to create a permanent human settlement in space or when will this happen. What can be predicted, however, is that other national space agencies and companies will rapidly follow suit, especially if there are economic benefits to be obtained from these settlements. With this notion that human settlements in space are bound to appear in the following decades, and that many of these settlements will be privately funded, how can the international community ensure that all fundamental human rights are protected in these environments?

4.3 Universal Declaration of Human Rights

The Universal Declaration of Human Rights (UDHR) is broadly considered the foundation for international human rights law. Adopted by the United Nations General Assembly in 1948, partly in response to the terrible crimes committed in World War II, it has served as an inspiration for lawmakers to draft legislation that protects the dignity of all humans.[10] With the advent of NewSpace and the possible drafting of a sixth international space treaty, a Mars Agreement, it is important to reflect on this foundational text and ensure that the treaty provides the necessary provisions to uphold it.

---

[6]Jan Woerner, 'Moon Village' (*The European Space Agency website, 2016*) https://www.esa.int/About_Us/Ministerial_Council_2016/Moon_Village. Accessed 12 September 2020.

[7]Namrata Goswami, 'China's Get-Rich Space Program' *The Diplomat* (28 February 2019) https://thediplomat.com/2019/02/chinas-get-rich-space-program/. Accessed 12 September 2020.

[8]Amanda Kooser, 'Elon Musk drops details for SpaceX Mars mega-colony' *CNET* (16 January 2020) https://www.cnet.com/news/elon-musk-drops-details-for-spacexs-million-person-mars-mega-colony/. Accessed 12 September 2020.

[9]Kari Paul, 'Blue Origin: Bezos company aims to take people to moon by 2024' *The Guardian* (9 May 2019) https://www.theguardian.com/technology/2019/may/09/blue-origin-jeff-bezos-moon-announcement. Accessed 12 September 2020.

[10]United Nations, 'Human Rights Law' (*United Nations website*) https://www.un.org/en/sections/universal-declaration/human-rights-law/index.html. Accessed 12 September 2020.

While all articles can be considered of equal importance, not all of them will be covered in this essay. Most of them are thankfully so ingrained in society that there is no reason to think they would not be respected in space settlements. Article 5, for example, condemns the use of torture and other inhuman treatments. The omission of this article from the discussion should not be interpreted as the endorsement of space torture by the author. Instead, it should be viewed as the trust that no nation would allow such behaviour to occur in space settlements in any way, shape, or form.

4.3.1 Article 9: Ban on Arbitrary Detention

Article 9 briefly states that "no one shall be subjected to arbitrary arrest, detention or exile".[11] This right is also enshrined in Articles 9 and 11 of the International Covenant on Civil and Political Rights (ICCPR).[12]

Human psychology is deeply altered by isolation and exposure to extreme environments.[13] On Earth, these adverse alterations can be seen on scientific missions that travel to the Antarctic for long periods.[14] On space, they can be seen in astronauts and cosmonauts that spend several months aboard the International Space Station, and which require teams of psychologists to help them throughout the mission.[15] In fact, NASA considers that the risk of adverse cognitive or behavioural conditions and psychiatric disorders for planetary missions which last above three years is "likely" and that it could develop into "lost time injuries" or "illness per OSHA criteria".[16,17]

Under such oppressive conditions, the possibility of mental illnesses appearing on the crew must be considered. If one of the inhabitants of a space settlement becomes

[11] Universal Declaration of Human Rights (10 Dec. 1948), U.N.G.A. Res. 217 A (III) (1948), Art. 9.

[12] International Covenant on Civil and Political Rights (New York, 16 Dec. 1966) 999 U.N.T.S. 171 and 1057 U.N.T.S. 407, entered into force 23 Mar. 1976 [the provisions of article 41 (Human Rights Committee) entered into force 28 Mar. 1979].

[13] J. I. Pagel and A. Choukèr, 'Effects of isolation and confinement on humans-implications for manned space explorations' (2016) Volume 120, Issue 12 Journal of Applied Physiology https://doi.org/10.1152/japplphysiol.00928.2015. Accessed 12 September 2020.

[14] Ron Roberts, 'Psychology at the end of the world' (2011) Volume 24, 22–25, The Psychologist https://thepsychologist.bps.org.uk/volume-24/edition-1/psychology-end-world. Accessed 12 September 2020.

[15] Robert Lewis, 'Behavioral Health' (*National Aeronautics and Space Administration website,* 7 August 2017) https://www.nasa.gov/content/behavioral-health. Accessed 12 September 2020.

[16] National Aeronautics and Space Administration, 'Risk of Adverse Cognitive or Behavioral Conditions and Psychiatric Disorders' (*National Aeronautics and Space Administration website,* 30 July 2020) https://humanresearchroadmap.nasa.gov/Risks/risk.aspx?i=99. Accessed 12 September 2020.

[17] National Aeronautics and Space Administration, 'Guidelines for Risk Management' (16 October 2017) https://www.nasa.gov/sites/default/files/atoms/files/s3001_guidelines_for_risk_management_-_ver_g_-_10-25-2017.pdf. Accessed 12 September 2020.

a danger to themselves, others, or the mission, how could the rest of the crew prevent them from causing any harm?

In an effort to resolve similar situations aboard aircraft, The Tokyo Convention grants special powers to the commander of the vessel. Specifically, they are given the right and responsibility to restraint passengers under the following grounds: "to protect the safety of the aircraft, or of persons or property therein; or to maintain good order and discipline on board; or to enable him to deliver such person to competent authorities."[18]

A similar protocol may be established for crews aboard spacecraft, which could by extension also mean space settlements. However, there is a fundamental difference between travel by aircraft and by spacecraft: the duration of the trips. The restraint of unruly passengers in planes is usually limited to only a few hours. If an inhabitant of a space settlement needs to be restraint, they may have to remain like that for months before transport back to Earth is available. Giving private companies the means to restrict the liberties of their employees for long periods without a proper legal framework to regulate it could mean the frequent violation of the ban on arbitrary detentions.

The Mars Agreement should permit the restraint of crew members aboard spacecraft (and space settlements) only when mission safety is in jeopardy. Moreover, all uses of restraint should be carefully recorded to enable the lawfulness of the restraint to be assessed. Private companies who are found to have wrongfully restrained settlers must be held responsible for their actions before the law (in contrast to Article 10 of The Tokyo Convention which exonerates aircraft commanders and the operators of the aircraft from any responsibility for the treatment undergone by restraint passengers).[19]

4.3.2 *Article 12: Right to Privacy*

Article 12 of the UDHR states: "No one shall be subjected to arbitrary or unlawful interference with his privacy, family, home or correspondence, nor to unlawful attacks on his honour and reputation."[20]

In today's terrestrial working life, digital tools such as smartphones or personal computers are used both privately and as professional tools.[21] Likely, this duality of use will also occur in space settlements because of its efficiency: carrying and maintaining redundant objects in space missions is costly and unnecessary.

[18]Convention On Offences And Certain Other Acts Committed On Board Aircraft [The Tokyo Convention] (Tokyo, 14 September 1963) *entered into force 4 December 1969*, Art. 6.

[19]Convention On Offences And Certain Other Acts Committed On Board Aircraft [The Tokyo Convention] (Tokyo, 14 September 1963) *entered into force 4 December 1969*, Art. 10.

[20]Universal Declaration of Human Rights (10 Dec. 1948), U.N.G.A. Res. 217 A (III) (1948), Art. 12.

[21]Calle Rosengren & Mikael Ottosson, 'Employee monitoring in a digital context' (2016) in 'Digital sociologies', 181–194.

Even if crew members were allowed to have both private and professional laptops or smartphones, they would probably still depend on the company for communication. Due to the long distances between space settlements and the terrestrial satellite infrastructure, a simple e-mail may take several hours to reach its destination.[22] To overcome this limitation, it is reasonable to assume that a local network for inter-crew communication will be developed, which will likely be installed and controlled by the private company.

Given that crew members will have to use some type of tool owned by the company for private purposes, it is essential to understand the conditions under which monitoring said tools could be considered in violation of their right to privacy. The European Court of Human Rights has issued several rulings in regard to Article 8 of the European Convention of Human Rights (which is equivalent in content to Article 12 of the UDHR: the right to privacy) that could shed some light over these conditions.[23]

In a 2007 ruling, the Court asserted the following:

> Accordingly, the Court considers that the collection and storage of personal information relating to the applicant's telephone, as well as to her e-mail and Internet usage, without her knowledge, amounted to an interference with her right to respect for her private life and correspondence within the meaning of Article 8.[24]

Similarly, in a ruling issued 10 years later, the Court determined that Article 8 had been violated by a company monitoring correspondence (in the form of instant messaging) of one of its employees because he was not given previous warning that the activity could be overseen.[25] In fact, the Court has established that one of the factors by which to assert possible violations of the right to privacy is "whether the employee has been notified clearly and in advance of the possibility that the employer might monitor correspondence and other communications, and of the implementation of such measures". Other factors include the extent of the monitoring and whether there are legitimate reasons to justify it.[26]

The Mars Agreement must include all the provisions necessary to guarantee space settlement inhabitants' access to free, confidential, and unfiltered communication. Private companies should create and provide their employees with non-monitored communication channels for private use, and should also clearly warn crew members when their professional conversations are being overseen.

---

[22]Robbie Gonzalez, 'How will space colonists access the Internet on Mars?' *Gizmodo* (6 June 2013) https://io9.gizmodo.com/how-will-space-colonists-access-the-internet-on-mars-511693210. Accessed 12 September 2020.

[23]European Convention for the Protection of Human Rights and Fundamental Freedoms (Rome, 4 Nov. 1950), 312 E.T.S. 5, as amended by Protocol No. 3, E.T.S. 45; Protocol No. 5, E.T.S. 55; Protocol No. 8, E.T.S. 118; and Protocol No. 11, E.T.S. 155; entered into force 3 Sept. 1953 (Protocol No. 3 on 21 Sept. 1970, Protocol No. 5 on 20 Dec. 1971, Protocol No. 8 on 1 Jan 1990, Protocol 11 on 11 Jan 1998).

[24]*Copland v. The United Kingdom* ECHR 2007-I no. 62617/00, Art. 8.

[25]*Barbulescu v. Romania* ECHR 2017 no. 61496/08.

[26]European Court of Human Rights, 'Guide on Article 8 of the European Convention on Human Rights' (30 April 2020), para. 572.

It is essential to protect the right to privacy because it represents a keystone for the adequate protection of other rights, such as the right to assembly or the right to a family.

4.3.3 *Article 15: Right to a Nationality*

In long term space missions like settlements, where men and women of all different sexual orientations are free to fraternize and create a community, sexual relationships are bound to occur. If the result of one of the mentioned relationships is a pregnancy that, fortunately, ends with the safe birth of a child, the first human will have been born outside of the Earth. It may also be, depending on the nationality of the parents (or lack thereof), the first stateless child to be born in space.

Article 24, Section 3 of the ICCPR, signed and ratified by the vast majority of nations on Earth, states that "[e]very child has the right to acquire a nationality".[27,28] Similarly, Article 7 Section 1 of the Convention on the Rights of the Child, equally accepted by a majority of states, declares that "[t]he child shall be registered immediately after birth and shall have the right from birth to a name, the right to acquire a nationality".[29,30] Finally, and despite enjoying less international support, Article 29 of the International Convention on the Protection of the Rights of All Migrant Workers and Members of Their Families states: "Each child of a migrant worker shall have the right to a name, to registration of birth and to a nationality".[31,32]

The international community, therefore, agrees on the premise that every child must be given a nationality at birth. Moreover, as declared by the UN Human Rights Committee, "[s]tates are required to adopt every appropriate measure, both internally and in cooperation with other States, to ensure that every child has a nationality when

[27] United Nations, 'Status of Treaties: International Covenant on Civil and Political Rights' https://treaties.un.org/Pages/ViewDetails.aspx?src=TREATY&mtdsg_no=IV-4&chapter=4. Accessed 12 September 2020.

[28] International Covenant on Civil and Political Rights (New York, 16 Dec. 1966) 999 U.N.T.S. 171 and 1057 U.N.T.S. 407, entered into force 23 Mar. 1976 [the provisions of article 41 (Human Rights Committee) entered into force 28 Mar. 1979], Art. 24.

[29] United Nations, 'Status of Treaties: Convention on the Rights of the Child' https://treaties.un.org/Pages/ViewDetails.aspx?src=TREATY&mtdsg_no=IV-11&chapter=4. accessed 12 September 2020.

[30] Convention on the Rights of the Child (New York, 20 Nov. 1989) 1577 U.N.T.S. 3, 28 I.L.M. 1448 (1989), entered into force 2 Sept. 1990, Art. 7.

[31] United Nations, 'Status of Treaties: International Convention on the Protection of the Rights of All Migrant Workers and Members of their Families' https://treaties.un.org/Pages/ViewDetails.aspx?src=TREATY&mtdsg_no=IV-13&chapter=4. Accessed 12 September 2020

[32] International Convention on the Protection of the Rights of All Migrant Workers and Members of their Families (New York, 18 Dec. 1990) 2220 U.N.T.S. 3 entered into force 1 July 2003, Art. 29.

he is born".[33] While this does not mean that states have the responsibility to grant citizenship to all children born in their territory, it can be interpreted as an obligation for states to grant their nationality as a "last resource" to prevent the child from being stateless.

This interpretation is supported by Article 1 Section 1 of the Convention on the Reduction of Statelessness, of which 75 nations are a party.[34] It asserts that "[c]ontracting States shall grant their nationality to persons, otherwise stateless, born in their territory".[35]

This statement brings us to a new point of discussion: the matter of territory in space. According to Article 2 of the Treaty on Principles Governing the Activities of States in the Exploration and Use of Outer Space, including the Moon and Other Celestial Bodies (Outer Space Treaty), "[o]uter space [...] is not subject to national appropriation by claim of sovereignty, by means of use or occupation, or by any other means."[36] The wording of this article leaves no room for interpretation: since no territory can be claimed in space, no nation can be made responsible for providing a nationality to stateless children.

We may turn again to the Convention on the Reduction of Statelessness, in this case to Article 3, for an answer to such a particular scenario: "For the purpose of determining the obligations of Contracting States under this Convention, birth on a ship or in an aircraft shall be deemed to have taken place in the territory of the State whose flag the ship flies or in the territory of the State in which the aircraft is registered, as the case may be".[37]

The Mars Agreement needs to add a similar provision to that of Article 3 of the Convention on the Reduction of Statelessness. All children born in space settlements that would otherwise be stateless must be given a nationality by the state that has jurisdiction over the settlement. For the case of private space settlements, the jurisdiction would belong to the state on whose registry the mission was launched, as per Article VIII of the Outer Space Treaty.[38]

---

[33] Human Rights Committee, General Comment 17, Art. 24 subdivision 8 (Thirty-fifth session, 1989).

[34] United Nations, 'Status of Treaties: Convention on the Reduction of Statelessness' https://treaties.un.org/pages/ViewDetails.aspx?src=TREATY&mtdsg_no=V-4&chapter=5. Accessed 12 September 2020.

[35] Convention on the Reduction of Statelessness (New York, 30 August 1961) 989 U.N.T.S. 175 entered into force 13 December 1975, Art. 1.

[36] Treaty on Principle Governing the Activities of States in the Exploration and Use of Outer Space, Including the Moon and Other Celestial Bodies [Outer Space Treaty] (Washington, 27 Jan. 1967), 610 U.N.T.S. 205; 18 U.S.T. 2410; T.I.A.S. No. 6347, entered into force 10 Oct. 1967, Art. 2.

[37] Convention on the Reduction of Statelessness (New York, 30 August 1961) 989 U.N.T.S. 175 entered into force 13 December 1975, Art. 3.

[38] Treaty on Principle Governing the Activities of States in the Exploration and Use of Outer Space, Including the Moon and Other Celestial Bodies [Outer Space Treaty] (Washington, 27 Jan. 1967), 610 U.N.T.S. 205; 18 U.S.T. 2410; T.I.A.S. No. 6347, entered into force 10 Oct. 1967, Art. 8.

4.3.4 *Article 16: Right to Marriage and Family*

Article 23 Section 2 of the ICCPR declares that "[t]he right of men and women of marriageable age to marry and to found a family shall be recognized".[39]

It is not uncommon for companies across the globe to forbid romantic relationships between employees. This prohibition is put in place to avoid issues such as gossip in the workplace, excessive distractions, damage to the reputation of the company, and even lawsuits.[40] Moreover, when a relationship forms between an employee and someone in a position of power, issues of favouritism may also arise.

Even though romantic relationships in the workplace can create conflicts of interest, and despite the rules put forth by employers that aim to prevent them, many people continue to find their partner at work. A survey of 5,795 UK workers done by Total jobs in 2018 revealed that 66% of the participants would be open to forming a relationship with a colleague and that 22% of them had met their partners at work (more than through friends and through online dating).[41]

Considering that in the early stages of privately-funded settlements a majority of the inhabitants will be employed by the founding company, it can be predicted that the majority of relationships would necessarily have to happen between employees. If the company were then to prohibit relationships between its employees, as it is sometimes done on Earth, would that be infringing upon their right to marriage and family?

There exist no relevant legal precedents on Earth because the conditions are unique to privately controlled settlements. Terrestrial social circles are not limited to colleagues, as they also commonly include neighbours, friends from school, roommates… Even if relationships are forbidden in the workplace, there are countless other opportunities to develop deeper relationships, and thus to retain the ability to form a family. Space settlers, on the other hand, may only be able to create close bonds with their colleagues.

However, the reasons why companies choose to ban romantic relationships at terrestrial workplaces cannot be dismissed: an efficient and fair work environment is equally if not more desirable in space.

The UN Humans Rights Committee also raised two more points related to the right to marriage and family in 1990. It was decided that "the right to found a family implies, in principle, the possibility to procreate and live together".[42]

---

[39] International Covenant on Civil and Political Rights (New York, 16 Dec. 1966) 999 U.N.T.S. 171 and 1057 U.N.T.S. 407, entered into force 23 Mar. 1976 [the provisions of article 41 (Human Rights Committee) entered into force 28 Mar. 1979], Art. 23.

[40] 'Ask An HR Expert: Romantic Relationships In The Workplace', (*TandemHR*, 14 February 2019) https://tandemhr.com/romantic-relationships-in-the-workplace/. Accessed 12 September 2020.

[41] 'How love works', (*Totaljobs*) https://www.totaljobs.com/insidejob/how-love-works/. Accessed 12 September 2020.

[42] Human Rights Committee, General Comment 19, Art. 23 subdivision 5 (Thirty-ninth session, 1990).

Let us consider the right to procreate. As it has been established above, settlers aboard privately-owned space outposts will likely only be able to have personal contact with their colleagues. By extension, these would be the only candidates with whom they might exercise their right to procreate. It follows then, that banning sexual relationships between employees would also be a violation of their right to marriage and family, even if non-sexual relationships were allowed.

It is important, however, to consider the consequences of sexual relationships happening on board. While several other important concerns could be discussed, such as the potential spread of sexually transmitted diseases, the focus in this section will be on issues related to pregnancy.

Permanent settlements will eventually have to be able to support pregnancies, but communities in early phases of development may not be able to so. Lack of medical equipment or of facilities capable of ensuring a safe birth could put the life of the mother and the child in danger. Furthermore, pregnancies that occur during interplanetary trips, which could take several months, could be under an even greater threat because of space radiation.[43]

The Mars Agreement must seek to protect the right of space settlers to found a family while giving provisions to companies to prevent dangerous behaviour in settlements. Romantic relationships between employees should only be banned when the relationship poses a reasonable danger to the mission or when one of those involved has authority over the other. Moreover, sexual relationships should only be prohibited for acts that could lead to a pregnancy, and only if the pregnancy could pose a danger for the child, the mother, or the mission. Finally, in recognition of the previously discussed UN Humans Rights Committee's interpretation of Article 23, the partners of a romantic relationship should be allowed to share living quarters.

4.3.5 Article 20: Right to Freedom of Assembly and Association

Article 20 Section 1 of the UDHR states that "[e]veryone has the right to freedom of peaceful assembly and association".[44] The focus of this section, however, will be a specific form of association: trade unions. These organizations are explicitly defended in Article 23 Section 4 of the UDHR, which declares that "[e]veryone has the right to form and to join trade unions for the protection of his interests", and in Article 22 Section 1 of the ICCPR: "Everyone shall have the right to freedom

[43]Marco De Santis and others, 'Radiation effects on development' (2007) Volume 81 Issue 3 Birth Defects Research Part C: Embryo Today 177–182 https://doi.org/10.1002/bdrc.20099. Accessed 12 September 2020.

[44]Universal Declaration of Human Rights (10 Dec. 1948), U.N.G.A. Res. 217 A (III) (1948), Art. 20.

of association with others, including the right to form and join trade unions for the protection of his interests."[45,46]

The advent of privately funded space settlements could mean the re-emergence of the company town model. Company towns were municipalities where almost every store and housing were owned by one company, which also acted as the employer of the population. Their founders wanted to create a better life for their employees and in return prevent their employees from "fall[ing] prey to the blandishments of union organizers".[47]

The efforts made by the company owners to prevent the unionization of the town inhabitants were usually successful. For instance, bosses of the coal-mine industry in the Northwest of the United States quickly evicted any workers who tried to organize, thus preventing unions from ever appearing. Even when companies tolerated and collaborated with unions, wages tended to be lower in those industries where company towns were common.[48]

The downfall of American company towns was brought by two phenomena that diminished the power that company owners had over their communities: legislative changes and an increase in the competitiveness of the labour market.[49] The founding corporation of the settlement will likely monopolize the labour market, making it considerably uncompetitive. Settlers will either have to work for a particular company or abandon the settlement, a process that could take months of travel and have prohibitive costs. It is therefore through legislation that excessive control by the company must be prevented and the right to unionize protected.

Let us now discuss the right to strike, both as a necessary tool for the complete fulfilment of the right to unionize and as an expression of the right to peaceful assembly. It is essential for trade unions to have tools to demonstrate their increased bargaining power as a collection of workers. Strike action represents one of such tools, and possibly the most important one. In the words of The Honourable Rosalie Silberman Abella, Puisne Justice of the Supreme Court of Canada, "along with their right to associate, [...] the right of employees to strike is vital to protecting the meaningful process of collective bargaining."[50]

Despite not being explicitly enshrined by the UDHR, the right to strike is recognized as a right and is regulated by legislation in many nations around the world,

---

[45]Universal Declaration of Human Rights (10 Dec. 1948), U.N.G.A. Res. 217 A (III) (1948), Art. 23.

[46]International Covenant on Civil and Political Rights (New York, 16 Dec. 1966) 999 U.N.T.S. 171 and 1057 U.N.T.S. 407, entered into force 23 Mar. 1976 [the provisions of article 41 (Human Rights Committee) entered into force 28 Mar. 1979], Art. 22.

[47]Linda Carlson, *Company Towns of the Pacific Northwest* (University of Washington Press, 2017) 187.

[48]Linda Carlson, *Company Towns of the Pacific Northwest* (University of Washington Press, 2017) 210.

[49]Linda Carlson, *Company Towns of the Pacific Northwest* (University of Washington Press, 2017) 211.

[50]Saskatchewan Federation of Labour v. Saskatchewan, 2015 SCC 4 (CanLII), [2015] 1 SCR 245, para. 24 http://canlii.ca/t/gg40r#par24. Accessed 12 September 2020.

including countries such as the United States of America,[51] Uruguay,[52] South Africa,[53] South Korea[54] and all member states of the Council of Europe (which is formed by European nations including Russia and Turkey).[55,56]

Even when the right to strike is not regulated, employees will still resort to striking to make their voices heard, such as in the case of the only strike to ever occur in space. In 1973, three astronauts, Jerry Carr, Ed Gibson, and William Pogue, took an unauthorized day off and cut the radio for 24 h after their complaints about their "16 h a day for 84 straight days" work schedule were repeatedly ignored by Mission Control.[57]

While this "incident" was resolved without any damage being done to the crew members or the mission, it demonstrated that even astronauts, some of the most tenacious and self-sacrificing professionals, are willing to defy direct orders if the working conditions become unacceptable. It also showed that the fact that striking rights are not protected or regulated does not mean that striking will not occur, only that it may occur in unexpected, uncontrolled, and potentially endagering conditions.

The right to unionize and to strike must therefore be included in any international agreement that hopes to ensure justice and fairness in private space settlements. Failure to set up a legal framework that provides safe and controlled mechanisms for employees to fight for a better quality of life could not only be a violation of their fundamental human rights but could also be a danger to space missions.

Perhaps the idea that the international community would agree to protect striking in space is too hopeful, maybe even too naïve. The mistakes of the Moon Treaty must not be repeated; it is essential to draft a Mars Agreement that all space-faring nations can abide. Still, this article makes the case for the right to strike because the author firmly believes that its protection is in the best interest of humankind as it strives to make space its home.

4.4 Conclusion

It is becoming increasingly clear that human settlements are a part of the future of the space sector. National space agencies around the world have displayed interest over the possibility of setting up permanent bases in the Moon or the Red Planet.

---

[51]National Labor Relations Act, 29 U.S.C. §§ 151–169 (1935).

[52]Constitution of Uruguay, (1997) Art. 57.

[53]Constitution of the Republic of South Africa, (1996) Art. 23.

[54]Constitution of the Republic of Korea, (1948) Art. 33.

[55]European Court of Human Rights, 'Guide on Article 11 of the European Convention on Human Rights' (31 May 2020), para. 254.

[56]Council of Europe, '47 Member States' (*Council of Europe website*) https://www.coe.int/en/web/portal/47-members-states. Accessed 12 September 2020.

[57]Michael Hiltzik, 'The day when three NASA astronauts staged a strike in space' *Los Angeles Times* (28 December 2015) https://www.latimes.com/business/hiltzik/la-fi-mh-that-day-three-nasa-astronauts-20151228-column.html. Accessed 12 September 2020.

Research is continuously being done on the extreme environments of outer space and how they could support long-term human occupation. Even the private sector, looking for lucrative ventures in space, has started drafting plans for extra-terrestrial human settlements.

The focus of this article is precisely on privately-funded settlements. More specifically, how the unique life conditions in these settlements could represent a challenge for the adequate protection of fundamental human rights, and the role of the Mars Agreement in that protection.

The ban on arbitrary detention (Article 9 of the UDHR) could be threatened by companies that abuse special powers given to them to ensure mission safety, for example by unnecessarily restraining crew members. The right to privacy (Article 12 of the UDHR), on the other hand, could be violated by companies overzealously monitoring communications channels not only when they are being used as professional tools, but also when used privately.

Moreover, to ensure the right to a nationality of every child (Article 15 of the UDHR), state parties to the treaty must be compelled to grant their citizenship to otherwise stateless children born in space settlements funded by companies under their jurisdiction.

The Mars Agreement should also prohibit unreasonable bans on relationships between crew members, thus protecting the right to marriage and family (Article 16 of the UDHR), while recognizing that some prohibitions are still necessary for the protection of the settlement and its members. Finally, the right to freedom of assembly and association (Article 20 of the UDHR) should be respected by protecting the formation of trade unions and the use of strike action in space, as long as mission safety is guaranteed.

Becoming an interplanetary specie has been the dream of science fiction authors and space enthusiasts for decades. However, we must remember that we have now the same excitement for expansion that colonizers of past centuries also had. These men and women made the terrible mistake of disregarding the value of human life in their quest for power and influence. It is our duty to draft a Mars Agreement that does not make space settlements simply a new manifestation of colonialism, but an enterprise that serves all humans equally. To quote John F. Kennedy, "space can be explored and mastered without feeding the fires of war, without repeating the mistakes that man has made in extending his writ around this globe of ours."[58]

Juan García Bonilla is an Aerospace Engineering student in the Universidad Carlos III de Madrid with a focus on space vehicles and doing research on the field of orbital mechanics. His passion for space makes him interested in areas other than engineering, such as space sustainability or space law.

[58] John Fitzgerald Kennedy, 'Address at Rice University on the Nation's Space Effort' (12 September 1962).

Chapter 5
Legal Implications of Detection of or Contact with Extra-Terrestrial Intelligence in the Mars Agreement Including Human Settlements

Tugrul Cakir

Abstract Detection of or contact with Extra Terrestrial Intelligence (ETI) is not beyond the realm of possibility during Mars colonisation. This potential aspect should be addressed in the Mars Agreement Including Human Settlements. Space treaties do not expressly deal with the issue. There is no formal international structure for addressing the issues of detecting, communicating or contacting ETI. Obligations of States stemming from space treaties should be clarified and strengthened. In this respect, Protocols formulated by the International Academy of Astronautics SETI Permanent Committee are a useful guide. Legal relationships between humankind and ETI are not likely to be treated in the Agreement but this issue should be revisited. The possibility of detection or the encounter with ETI makes space lawyers consider our legal relationships with ETI.

5.1 Introduction

In the near future, new space activities will cause space lawyers think about the relevance of space treaties, which were not prepared as a response to the needs of practice. In less than two decades, space colonisation will no longer be a dream.[1] Indeed, the realisation of colonisation is the last step of human activities taking place in a space (after exploration and utilisation). Mars is the most privileged destination of space colonisation thanks to its favourable characteristics. The colonisation project could be conceived as an international collective effort. National laws of their State of Registry would apply to Mars settlers but an agreement among participating States to the joint international project establishing an international legal framework is also necessary. One of the goals of the Mars Agreement, including human settlements

[1]For a discussion on the necessity of Mars colonisation see generally Igor Levchenko, Shuyan Xu, Stéphane Mazouffre, Michael Keidar and Kateryna Bazaka, "Mars Colonization: Beyond Getting There" (2019) 3 Global Challenges 1. https://onlinelibrary.wiley.com/doi/10.1002/gch2.201970011. Accessed 25 August 2020.

T. Cakir (✉)
Faculty of Law, Ankara Yildirim Beyazit University, Ankara, Turkey
e-mail: tcakir@ybu.edu.tr

A. Froehlich (ed.), *Assessing a Mars Agreement Including Human Settlements*, Studies in Space Policy 30, https://doi.org/10.1007/978-3-030-65013-1_5

would be to prevent conflicting and competing jurisdictions especially in the area of penal law and intellectual property law as in the case of the Agreement on the International Space Station.[2]

One of the other objectives of the Mars Agreement would be to promote and follow a joint approach in issues such as the possibility of discovering and coming into contact with Extra-Terrestrial Intelligence (ETI). This text will focus only on the discovery of and contact with intelligent extraterritorial life. Excluded from the scope of this study is the discovery of and the contact with non-intelligent extra-terrestrial life.[3] For the purposes of this discussion, "intelligence" is defined as "the ability to learn, understand and think in a logical way about things; the ability to do this well".[4]

The question of the existence or the absence of extra-terrestrial intelligence has attracted attention for centuries. We have not yet discovered nor had contact with ETI since the beginning of space exploration.[5] Even so, we have only been searching for Extra-Terrestrial Intelligence (SETI) only since the 1960s.[6] This is, however, a short period in the history of our galaxy and universe. The development of technology regarding exploration of our own solar system and the universe would permit us to learn more about the existence of ETI.

A detection of, or an encounter with, ETI could have psychological, social, religious, cultural and political implications on human societies.[7] Such a detection could worsen conflicts between humans, especially on how messaging to ETI (METI) should occur, what we should say or who speaks on the behalf of earth.[8] The location of contact, which could be on earth, within the orbit of earth or beyond earth's orbit, could have consequences on the intensity of reactions of the people of earth.[9] The final scenario (a contact beyond the realm of earth's orbit) would seem likely to be the less "dramatic", but a possible contact with settlers on Mars

---

[2]Agreement among the government of Canada, governments of member states of the European Space Agency, the government of Japan, the government of the Russian Federation, and the government of the United States of America concerning cooperation on the civil international space station, signed in Washington 29 January 1998, entry into force 27 March 2001.

[3]On the discovery of non-intelligent extraterrestrial life see generally Margaret S. Race and Richard O. Randolph, "The need for operating guidelines and a decision framework applicable to the discovery of non-intelligent extraterrestrial life" (2002) 30 Advances in Space Research 1583, 1591. https://www.sciencedirect.com/science/article/pii/S0273117702004787. Accessed 25 August 2020

[4]Oxford Dictionaries. https://www.oxfordlearnersdictionaries.com/definition/english/intelligence?q=intelligence. Accessed 25 August 2020.

[5]Philippe Ailleris, "UFOs and Exogenous Intelligence Encounters" (2011) 43 ESPI Perspectives 11. https://espi.or.at/search-results?q=UFOs%20and%20Exogenous%20Intelligence%20Encounters. Accessed 25 August 2020.

[6]On SETI programmes see generally Francis Lyall and Paul B. Larsen, *Space Law: A treatise* (Ashgate 2009) 540, 543.

[7]Ailleris (no 5) 3; Seth D. Baum, Jacop Haqq-Misra and Shawn Domagal-Goldman, "Would contact with extraterrestrials benefit or harm humanity? A scenario analysis" (2011) 68 Acta Astronautica 3.1.

[8]Ibid.

[9]Michael T. Schetsche, "SETI (Search for Extraterrestrial Intelligence) and the Consequences: Futurological Reflections on the Confrontation of Mankind with an Extraterrestrial Civilization"

would change things completely. Indeed, "Space colonisation will increase the probability of discovering and coming into contact with extraterrestrial intelligence, either biological or artificial".[10]

Space treaties do not expressly deal with this issue. There is no formal international structure for addressing the issues of detecting, communicating or contacting ETI. The issue may be found to be unnecessary or "sciencefictioneerish"[11] because the chances of discovering ETI is quite low. Nevertheless, it is not necessary to wait for contact with ETI before considering addressing this issue. We should prepare the legal means for the possibility of the first contact on Mars (both transmitting such a discovery and replying to such a contact). Elaborating a treaty entirely dedicated to the search for, contact, and communication with, ETI is premature.[12] Nevertheless, State Parties to the Mars Agreement—probably also engaged in SETI and METI activities—would have a great deal of interest in including a provision dealing with the issue.

In case of such a detection, the process should be coordinated by the International Mars Authority (IMA) established by participating States in order to promote cooperation in the peaceful exploration and use of Mars and to ensure the coexistence of applicable national laws. IMA shall cooperate with the United Nations and the International Academy of Astronautics SETI Permanent Committee in order to follow a collective approach. In terms of a reply to ETI, States are not under any obligation; so, they are free as to whether or not to reply.[13] Nevertheless, before sending a message to ETI, a worldwide scientific, political and humanitarian discussion would be appropriate because this would have consequences for all of the nations in the world.[14] A collective response should be drafted in a coordinated manner even if "it could be slow and laborious".[15]

The first contact would be origin of our legal relationship with ETI. It is evident that aliens are not addressees of "law" as such, including "space law".[16] The latter has

---

(*Astrosociology*, 1 July 2005) 3,6. http://www.astrosociology.com/Library/PDF/Contributions/SETIandConsequences_ENG.pdf. Accessed 25 August 2020.

[10]Marko Kovic, "Political, moral, and security challenges of space colonization" (Zipar, 12 June 2018) 4.2. https://zipar.org/discussion-paper/political-moral-security-challenges-space-colonization/. Last accessed 25 August 2020.

[11]Ernst Fasan, *Relations with Alien Intelligences—The Scientific Basis of Metalaw,* in Patricia Margaret Sterns and Leslie I. Tennen (eds) *Private Law, Public Law, Metalaw and Public Policy in Space A Liber Amicorum in Honor of Ernst Fasan* (Springer 2016) 189.

[12]Frans G. von der Dunk, "Shaking the Foundations of the Law: Some Legal Issues Posed by a Detection of Extra-Terrestrial Life" in James S.J. Schwartz and Tony Milligan (eds.), *The ethics of Space Exploration* (Springer 2016) 253. Bilder thinks that a United Nations General Assembly resolution or declaration of principles relating to SETI and METI issues would be more appropriate than elaborating a treaty. Richard B. Bilder, "On the search for extraterrestrial intelligence (SETI)" (2020) 114 American Journal of International Law 94, 95.

[13]Michael A.G. Michaud, "Ten decisions that could shake the world" (2003) 19 Space Policy 133, 134.

[14]Bilder (no 12) 93, 94.

[15]Michaud (no 13) 134.

[16]Von der Dunk (no 12) 257.

an anthropocentric character. Haley defines "anthropocentric law" as "being simply the law of human beings".[17] One can advance that "the entire legal perspective of the universe [is] completely geocentric, with Earth as the centre point of the legal universe".[18] Certain provisions of space treaties (*province of all mankind*[19] in Art. I of the Outer Space Treaty (OST)[20] and *the common heritage of mankind* in Art. 11 of the Moon Agreement (MOON)[21]) are "*geo-centric* by nature" and may be origin of a conflict between human beings and aliens.[22] It is necessary to underscore that International Space Law governs only space activities conducted by human beings. The legal relationship between humankind and ETI (so the behaviour of both parties) would be governed by mutual accords ensuring peace and stability.

Before delving into the matter of our legal relationships with ETI (3), it is necessary to recall the place of SETI and METI activities in International Space Law (2).

5.2 International Legal Aspects of SETI/METI

5.2.1 SETI in Binding Instruments

SETI activities are lawful under International Law and can be seen as a part of the principle of freedom of exploration of outer space[23] enshrined in Art. I OST.[24] The beneficiary of this freedom is mainly States and international intergovernmental organisations.[25] This freedom is also granted to non-governmental entities. According to Art. VI OST, States are responsible for their national activities, which include not only governmental ones but also the activities of non-governmental entities. State Parties shall also authorize and continuously supervise the activities of these entities.

---

[17] Andrew Haley, "Space Law and Metalaw—Jurisdiction Defined" (1957) 24 J. Air L. & Com. 286.

[18] Zach Miller, "Space settlement and the celestial subjectivity model: Shifting our legal perspective of the Universe" in Annette Froehlich (ed.), *A Fresh View on the Outer Space Treaty*, Springer, Studies in Space Policy, Vol. 13, 2018, p. 62.

[19] The term "mankind" is defined by Fasan as "the notion for the whole of all human beings, the whole of humanity". See generally Ernst Fasan, "The meaning of the term "Mankind" in space legal language" (1974) 2 Journal of Space Law 125, 131.

[20] Treaty on Principles Governing the Activities of States in the Exploration and Use of Outer Space, including the Moon and Other Celestial Bodies (Outer Space Treaty), adopted by the General Assembly in resolution 2222 (XXI), opened for signature on 27 January 1967, entered into force on 10 October 1967.

[21] Agreement Governing the Activities of States on the Moon and Other Celestial Bodies adopted 5 December 1979, opened for signature 18 December 1979, entered into force 11 July 1984.

[22] H. Peter van Fenema, "Space law" (1988) 13 Air Law 86.

[23] On the freedom of exploration and use see generally Stephan Hobe, "Article I", in Stephan Hobe, Bernhard Schmidt-Tedd and Kai-Uwe Schrogl (eds.) *Cologne Commentary on Space Law: Volume I, Outer Space Treaty* (Carl Heymanns Verlag 2009) 34, 35.

[24] Lyall and Larsen (no 6) 544.

[25] Hobe (no 23) 33.

It should be noted that international obligations stemming from space treaties are not binding for private entities, but States shall make sure that these entities comply with their obligations. Therefore, even if a SETI activity is conducted by a non-governmental entity, States must make sure that this activity is in accordance with International Law.

According to Art. I para. 3 of OST, States are also beneficiaries of the freedom of scientific investigation, which is "the most specific activity covered by "exploration"".[26] In such scientific investigation, States "shall facilitate and encourage international co-operation".[27] In line with Art. I, which incentivizes States to cooperate, Art. XI[28] provides that States Parties "agree to inform the Secretary-General of the United Nations as well as the public and the international scientific community, to the greatest extent feasible and practicable, of the nature, conduct, locations and results of [their space] activities. On receiving the said information, the Secretary-General of the United Nations should be prepared to disseminate it immediately and effectively." Indeed, this Article "appears as a remarkable development in setting up an information system based on the pro-activeness of the State Parties and the focal role of the UN Secretary General".[29]

Therefore, in case of any detection of an alien signal, the UN Secretary General shall be informed by a State Party.[30] Nevertheless, States shall disclose information concerning the nature, conduct, locations and results of their space activities "to the greatest extent feasible and practicable". Therefore, "The vague wording of this provision opens the floodgates to justifications not to disseminate information related to space activities".[31] Thus, a State may not disclose any detection of an alien signal privileging its strategic or political interests.[32] It would be suitable that the duty of information enshrined in Art. XI OST should be strengthened by a provision of the Mars Agreement in case of any indication of ETI.

Regarding potential danger to human life, there are two main articles in space treaties. Art. V OST provides that "States Parties to the Treaty shall immediately inform the other States Parties to the Treaty or the Secretary-General of the United

---

[26]Ibid 36.

[27]The cooperation principle is included in the following Articles of the OST: I, III, IX, X, XI. Indeed, the nature of space activities requires that Space Law is a law of cooperation. Today, the number of international organisations involved in space is even more so after the intensification and diversification of space activities.

[28]See generally Jean-François Mayence and Thomas Reuter, "Article XI", in Stephan Hobe, Bernhard Schmidt-Tedd and Kai-Uwe Schrogl (eds.) *Cologne Commentary on Space Law: Volume I, Outer Space Treaty* (Carl Heymanns Verlag 2009) 189, 206.

[29]Ibid 192.

[30]"The information is at the same time a *means* of co-operation and an *object* of co-operation. A *means*, because information on the activities and their modalities (nature, conduct, locations) aims to foster participation of other countries therein or contribution thereto; an *object*, because the result of the activities are at stake and constitute valuable resources for the co-operating countries". Ibid 191.

[31]Ibid 197.

[32]Patricia M. Sterns, "SETI and Space Law: Jurisprudential and Philosophical considerations for humankind in relation to extraterrestrial life" (2000) 46 Acta Astronautica 759, 760.

Nations of any phenomena they discover in outer space, including the Moon and other celestial bodies, which could constitute a danger to the life or health of astronauts." The appreciation of this type of phenomena rests with the States Parties, which are mainly addressees of obligation to provide information in such cases.[33] Indeed, this Article is an example of humanitarian character of the OST.[34] Nevertheless, this Article deals only with potential danger to life of astronauts but all the people living life on Mars settlement would not be considered as astronauts.[35] It is important to note that according to Art. 10 para. 1 MOON, "any person on the Moon" is to be regarded as an astronaut within the meaning of the Art. V OST.[36]

Regarding potential danger to human life, the other article is Art. 5 para. 3 MOON, which provides that "In carrying out activities under this Agreement, States Parties shall promptly inform the Secretary-General, as well as the public and the international scientific community, of any phenomena they discover in outer space, including the Moon, which could endanger human life or health, as well as of any indication of organic life." It is necessary to underscore that if there is no danger to human life or health endangered by a discovery, this paragraph is not applicable. It is also easy to note that the reference has been made to organic life in outer space not to extra-terrestrial intelligence. Nevertheless, Lyall and Larsen observe that "detection of or contact with ETI could be covered under a broader interpretation of "dangerous phenomena", "dangerous" at least *in potentia*".[37] It is important to remind that the MOON does not have major support from the space-faring States, which would also be a party of the Mars international collective effort. Therefore, States should be required, in the Mars Agreement, to promptly inform the International Mars Authority about dangers to the life and health of all settlers and about any indication of ETI. On receiving the said information, the Authority shall take appropriate measures to protect the life or health of settlers and inform the Secretary-General of the United Nations as well as the public and the international scientific community.

---

[33] Frans von der Dunk and Gérardine Meishane Goh, "Article V", in Stephan Hobe, Bernhard Schmidt-Tedd and Kai-Uwe Schrogl (eds) *Cologne Commentary on Space Law: Volume I, Outer Space Treaty* (Carl Heymanns Verlag 2009) 101.

[34] Bin Cheng, *Studies in international space law* (Clarendon Press 1997) 460.

[35] Von der Dunk and Goh (no 33) 96, 98.

[36] See generally Ram Jakhu, "Article 10 MOON", in Stephan Hobe, Bernhard Schmidt-Tedd and Kai-Uwe Schrogl (eds) *Cologne Commentary on Space Law: Volume II* (Carl Heymanns Verlag 2013) 385, 387. Moon Treaty is also applicable to all celestial bodies within the solar system including Mars "except insofar as specific legal norms enter into force with respect to any of these celestial bodies". On the area of application of MOON see generally Stephan Hobe and Fabio Tronchetti, "Article 1 MOON", in Stephan Hobe, Bernhard Schmidt-Tedd and Kai-Uwe Schrogl (eds.) *Cologne Commentary on Space Law: Volume II* (Carl Heymanns Verlag 2013) 351, 354.

[37] Lyall and Larsen (no 6) 551.

5.2.2 SETI/METI in Non-binding Instruments

There are two protocols in which the detection of or the reply to ETI have been directly addressed. These have been formulated by the International Academy of Astronautics SETI Permanent Committee, which has a mission "to provide a forum for the SETI scientists from all over the world where they can meet and discuss progress in the field of SETI".[38] The first is "the Declaration of Principles for Activities Following the Detection of Extraterrestrial Intelligence" (The Post-Detection Protocol), which was adopted in 1989 and revised in 2010.[39] The aim of this protocol is "to declare our commitment to conduct this search in a scientifically valid and transparent manner and to establish uniform procedures for the announcement of a confirmed SETI detection".[40] According to the first principle, SETI activities will be conducted transparently. The discoverer should make all verification efforts of a suspected detection (Pr. 2).[41] If a detection of a signal is confirmed, the discoverer shall report this conclusion to the public, the scientific community, and the Secretary General of the United Nations in compliance with OST Art. XI (Pr. 3). International consultations are necessary before replying to a confirmed detected signal (Pr. 8).

The second Protocol, "Declaration of Principles Concerning Sending Communications to Extraterrestrial Intelligence" (The Reply-communication Protocol) details procedural aspects of communication to extra-terrestrial intelligence.[42] The question of sending communications should be considered by consultations (Pr. 1). Indeed, international consultations have been conceived as a necessary step before sending a communication (Pr. 8). Consultations should take place mainly under the auspices of the Committee on the Peaceful Uses of Outer Space of the United Nations (UNCOPUOS) (Pr. 2). They should be open to all interested States (Pr. 3). Based on recommendations from the UNCOPUOS, United Nations General Assembly takes the final decision as to whether or not to reply (Pr. 4). If a decision be made to send a message, "it should be sent on behalf of all Humankind, rather than from individual States" (Pr. 5). The content of the message "should reflect a careful concern for the

---

[38]On the role played by the IAA SETI Permanent Committee in SETI see generally Claudio Maccone, "SETI and the IAA SETI Permanent Committee: Past, Present and Possible Future" in Patricia Margaret Sterns and Leslie I. Tennen (eds) *Private Law, Public Law, Metalaw and Public Policy in Space A Liber Amicorum in Honor of Ernst Fasan* (Springer 2016) 153, 158.

[39]IAA SETI Permanent Committee, "Declaration of Principles for Activities Following the Detection of Extraterrestrial Intelligence". https://iaaseti.org/en/protocols/. Last accessed 25 August 2020.

[40]Ibid Preamble.

[41]Patricia Margaret Sterns and Leslie I. Tennen observe that these protocols ignore the growth of social media and its implications and propose that protocols should be updated to this changing context in the age of social media. Patricia Margaret Sterns and Leslie I. Tennen, "SETI, Metalaw, and Social Media" in Patricia Margaret Sterns and Leslie I. Tennen (eds) *Private Law, Public Law, Metalaw and Public Policy in Space A Liber Amicorum in Honor of Ernst Fasan* (Springer 2016) 159, 179.

[42]IAA SETI Permanent Committee, "Draft Declaration of Principles Concerning Sending Communications to Extraterrestrial Intelligence". https://iaaseti.org/en/seti-reply-protocols/. Last accessed 25 August 2020.

broad interests and well-being of Humanity, and should be made available to the public in advance of transmission" (Pr. 6).

Both protocols are not legally binding for States. The first contains a set of principles regarding disseminating information about the detection of ETI to be followed by institutions involved in SETI. As advanced by Reynolds, "it does not necessarily lose importance, or likelihood of being followed, for that reason. Indeed, agreements among scientific researchers have a strong history of being followed and this one is likely to be looked to should the event occur."[43] The second is rather a draft and so has "no real status".[44]

Indeed, it is easy to point out the augmentation of number of soft law instruments in Space Law, because of their flexibility in comparison with other classical sources of Space Law. Even if these instruments are not binding, they may constitute a basis for a binding instrument and contribute to the crystallization of a customary rule, as was the case at the beginning of the conquest of space. In this context, SETI protocols should be a useful guide when clarifying obligations of States stemming from space treaties.

The nature of detection would have implications on the content of the reply to ETI.[45] The question here, "is not whether intelligent life exists elsewhere, but what to do if we contact it".[46] In this respect, the issue of legal relationship between humankind and ETI should be revisited.

5.3 From Anthropocentric Law to Meta Law

Haley, founder of Meta law, was the first author to deal with the legal aspects between sapient beings in the universe. According to the author, the anthropocentric concepts of law should not be applicable to our interaction with ETI. Haley proposes the Golden Rule, which is also the premise of Meta law: "We must do unto others as they would have done unto them. To treat others as we would desire to be treated might well mean their destruction. We must treat them as they desire to be treated."[47] In Haley's conception, rules will be established as part of a process of natural evolution based upon certain fundamental moral principles in case of an interaction with ETI.[48]

Haley's ideas were developed by Fasan, who referred to a moral obligation called by Immanuel Kant as the categorical imperative: "Act in such a way, that the maxim

[43] Glenn H. Reynolds, "International Space Law: Into the Twenty-First Century" (1992) 25 Vanderbilt Journal of Transnational Law 250. https://ssrn.com/abstract=2571912. Accessed 25 August 2020.

[44] Lyall and Larsen (no 6) 555.

[45] Patricia M. Sterns (no 32) 761.

[46] Reynolds (no 43) 246.

[47] Andrew Haley, *Space law and government* (Appleton Century Crofts 1963) 395.

[48] See generally Steven Freeland, "A Natural System of Law—Andrew Haley and the International Legal Regulation of Outer Space" (2013) 39 Journal of Space Law 77, 98.

of your will can at the same time always be valid as the principle of general legislation".[49] This formula is applicable to all intelligent beings with five characteristics in common: life, intelligence, detectability (by the other race), three-dimensionality, the will to live.[50] Thus, if a certain action would result in a contradiction or a destructive result as a general rule, it is forbidden by the imperative deducing from the concept of reason alone.

For Fasan, "Metalaw is a new kind of natural law whereby the "nature" upon which this law is based is the mutual biological characteristics of all intelligent races".[51] This approach is a necessary step, for the author, in order to evolve the rules of Meta law. From there, Fasan elaborates 11 fundamental rules of Meta law: the strongest one is "*no partner of Metalaw may demand an impossibility*".[52] After clarifying these positive norms, Fasan refers to the Pure Theory of Law of Hans Kelsen. Fasan thinks that it is essential in order to elaborate new Meta legal norms.[53] The legal system created by Kelsen, which is "only dependent on the basic notion of intelligent life", "will provide us with enough knowledge of theoretical formal legal structure to enable us to formulate any legal notions of *positive law desired* in the necessary *structure of law*".[54] According to Fasan, new legal notions (or rules of behaviour) would be formulated as part of a cooperation between intelligent races.[55]

The goal of elaborating Meta legal rules in compliance with the Natural Law Theory is to avoid mutually harmful interactions in case of a contact between human beings and ETI.[56] However, it is not possible to guarantee that humankind would abide by these rules in its relationships with ETI given human history and experience.[57] There is also no guarantee that these rules would be respected by ETI. As put forward by Robinson, the cultural concept of rules (including the Golden Rule of Haley) is also anthropocentric; the determination of "which is 'injurious or hurtful to some other being?'" is made with anthropocentric considerations.[58] Therefore, applying the Golden Rule in our dealings with ETI may be problematic.

The nature of legal relationships between humankind and ETI would take shape depending upon the level of intelligence and advancement of extra-terrestrial life. Von der Dunk foresees three ETI contact scenarios: (1) extra-terrestrial life less intelligent and advanced than human life, (2) as intelligent and advanced as human

49 Fasan (no 11) 225.

50 Ibid 226.

51 Ibid 229, 230.

52 Ibid 230, 238.

53 Ibid 238.

54 Ibid.

55 Ernst Fasan, "Legal consequences of SETI detection" (1998) 42 Acta Astronautica 678.

56 George S. Robinson, "Metalaw: From Speculation to Humankind Legal Posturing with Extraterrestrial Life" (2013) 2 Journal of Space Philosophy 50.

57 Sterns and Tennen (no 41) 164.

58 George S. Robinson, "Ecological Foundation of Haley's Metalaw" (1969) 22 J. British Interplanetary Soc. 266, 274 as cited in Robinson (no 56) 49, 50.

life, (3) much more intelligent and advanced than human life.[59] In the first case, ETI probably has no concept of law and the manner of their treatment by humans could be decisive.[60] In the second case, they have likely have a concept similar to law; finding the means to forge a compromise system or construct for inter-species social would be a possible option.[61] In the last case, ETI may be benevolent (updating its own goals and preferences in the light of that of humankind), apathetic (not updating its own goals and preferences in the light of that of humankind) or hostile.[62] An eagerness to consume our resources and a desire to be the sole galactic power may be origin of hostility of ETI to humanity.[63] ETI may also be unintentionally harmful to humanity such as by the transmission of disease to humanity.[64] Finally, it can be advanced that "if ETI are significantly more advanced than humanity, then the outcome of contact may depend primarily on ETI desires".[65]

Establishing of Meta law rules has a prospective character and they would be established after the first interaction between humankind and ETI. "*Without knowing anything about the different intelligent races of the universe, we cannot therefore definitely establish rules of Meta law.*"[66] Direct contact with ETI would lead us to translate the principles of Meta law into positive law, especially in the agreements concluded governing mutual relations between ETIs and human beings, such as "galactic protocols". These principles, which are immanent in nature and in conformity with reason and ensure social peace, belong to the moral sphere and are likely to guide the legal basis of our relations with ETI. This relationship would confirm a paradigm shift from an anthropocentric law to Meta law. The first would not expand to include the second but they should coexist. Human laws should be applicable only for mankind. The agreements concluded after the first interaction should be respected by both parties.

5.4 Conclusion

According to Bin Cheng, there are three main conditions for successful treaty making in Space Law: perceived need, propitious climate and due representation of the interests during the law-making process.[67] These conditions regarding the Mars Agreement would be satisfied in the future. A provision dealing with SETI/METI issues should be addressed in the Mars Agreement. A detection of or a contact with ETI

[59] Von der Dunk (no 12) 262, 263.

[60] Ibid 262.

[61] Ibid 262, 263.

[62] Kovic (no 10).

[63] Baum, Haqq-Misra and Domagal-Goldman (no 7) 5.

[64] Ibid 6.

[65] Ibid 2.4.

[66] Fasan (no 11) 226.

[67] Cheng (no 34) 687, 694.

on Mars would have important impacts on human societies. Thus, we should be prepared by legal means for the possibility of the first contact on Mars. It would be suitable to include the following article in the Mars Agreement:

> **Proposed Article** In carrying out activities under this Agreement, States Parties shall promptly inform the International Mars Authority, of any phenomena they discover in outer space, including the Mars, which could endanger human life or health of settlers, as well as of any indication of extra-terrestrial intelligence.
> On receiving the said information, the Authority shall take appropriate measures to protect the life or health of settlers and inform the Secretary-General of the United Nations as well as the public and the international scientific community.
> In the event of a suspected detection of extra-terrestrial intelligence, the Authority shall make all efforts to verify the detection.
> The Authority shall undertake appropriate international consultations with the United Nations and the SETI Committee of the International Academy of Astronautics before proceeding with any reply to an extra-terrestrial intelligence.
> Detailed provisions in implementation of this Article shall be made in a Memorandum of Understanding between participating States.

Tugrul Cakir is a lecturer specialised in Space Law at the Ankara Yildirim Beyazit University. He earned a master's degree and a Ph.D. degree at the Université Jean Moulin Lyon III (France). His master's thesis and Ph.D. thesis treated different aspects of International and National Space Law.

Chapter 6
The International Space Station (ISS) Intergovernmental Agreement as a Precedent for Regulating the First Human Settlements on Mars

Alexandros Eleftherios Farsaris

Abstract In spite of the global crisis the world is suffering from, the space sector has not stopped evolving. The rise of private space actors and the increasing commercialisation of outer space, further incentivise the development and growth of new technologies that enhance space exploration. With the current rate of scientific and technical development, the reality of establishing the first human settlements on Mars is well within the foreseeable future. Legal questions arising from the establishment of such settlements and from the human activities thereon, will be a pressing issue since the general regime provided by international space law does not offer sufficient regulations. In providing for a new international regime, an important precedent in the hands of the legislators will be the International Space Station (ISS) legal framework. The ISS' proven legal regime, and especially the Intergovernmental Agreement, can play a significant role in providing solutions for a number of issues, such as criminal jurisdiction, liability, intellectual property and the regulation of outer space commercialisation.

6.1 Introduction

The colonization of Mars is an inherently fascinating subject and along the years it has received interest from space agencies and private corporations, while it has also received extensive treatment in science fiction, film and art. Even if the idea of colonizing Mars has been around for many decades, it was always in the realm of science fiction rather than in the realm of reality. However, as of 2020, SpaceX, a private aerospace manufacturer and space transportation services company, is testing prototypes of the 'Starship', the company's designated spacecraft for cargo and crew

[1]Mike Wall, 'SpaceX gearing up for 12-mile-high test flight with prototype of Mars-colonizing Starship' (LIVESCIENCE 15 September 2020). www.livescience.com/amp/spacex-starship-sn8-prototype-test-flight-soon.html. Last accessed 15 September 2020.

A. E. Farsaris (✉)
University of Luxembourg, Esch-sur-Alzette, Luxembourg
e-mail: alex.farsaris@gmail.com

A. Froehlich (ed.), *Assessing a Mars Agreement Including Human Settlements*, Studies in Space Policy 30, https://doi.org/10.1007/978-3-030-65013-1_6

transportation to and from Mars.[1] Even if the development of such spacecraft is at its early stages, the first flights transporting equipment and crew are scheduled to land on Mars well within this decade.[2] Moreover, following the success of the first missions, SpaceX estimates to establish the first 'Mars cities' in the next few decades.[3] Although it can be argued that some of these estimates are rather optimistic, it is clear that landing the first humans on Mars, and eventually establishing the first bases, is slowly exiting the realm of science fiction, and entering the realm of reality. Among the numerous scientific and technical issues, establishing human settlements on Mars would also face a number of legal challenges.

When dealing with outer space, the community of states decided that international law applies to activities in space, leading to the establishment of five United Nations (UN) outer space treaties.[4] The UN treaties form the core body of space law, the legal basis upon which all outer space activities are based. However, they mostly provide general principles which albeit being essentially the 'corpus juris spatialis internationalis', they do not regulate complex situations that were not foreseen in the days when they were enacted.

A human establishment on Mars, given the sheer scale of such an endeavour, would most likely require a global effort involving states and private entities, demanding specific provisions for the regulation of the relations deriving from such a cooperation. Hence, an agreement regulating the human establishment on Mars would have to provide for fields for which the UN treaties establish a limited regime, or do not establish a regime at all, such as criminal jurisdiction, the protection of intellectual property rights and the involvement of the commercial sector in outer space activities.

Nevertheless, establishing a long-term extra-terrestrial outpost on Mars would not come unprecedented.

---

[2]Meghan Bartles, 'SpaceX wants to send people to Mars. Here's what the trip might look like.' (Space 26 May 2020). www.space.com/spacex-plans-journey-to-mars.html. Last accessed 15 September 2020.

[3]Darrell Etherington, 'Elon Musk says building the first sustainable city on Mars will take 1000 Starships and 20 years' (Techcrunch 8 November 2019). https://techcrunch.com/2019/11/07/elon-musk-says-building-the-first-sustainable-city-on-mars-will-take-1000-starships-and-20-years/. Last accessed 15 September 2020.

[4]Treaty on Principles Governing the Activities of States in the Exploration and Use of Outer Space, including the Moon and other Celestial Bodies, 19 December 1966 (entered into force 10 October 1967) 610 UNTS 205 (Outer Space Treaty); Agreement on the Rescue of Astronauts, the Return of Astronauts and the Return of Objects Launched into Outer Space, 19 December 1967 (entered into force 3 December 1968), UNGA Res. 2345 (XXII) (Rescue Agreement); Convention on International Liability for Damage Caused by Space Objects (adopted 29 November 1971, entered into force 1 September 1972) UNGA Res 2777 (XXVI) (Liability Convention); Convention on Registration of Objects Launched into Outer Space (adopted 12 November 1974, entered into force 15 September 1976) UNGA res 3235 (XXIX) (Registration Convention); Agreement Governing the Activities of States on the Moon and Other Celestial Bodies (adopted 5 December 1979, entered in force July 1984) UNGA res 34/68 (Moon Agreement).

6.1.1 The International Space Station and the Intergovernmental Agreement

The International Space Station (ISS) is a cooperative scientific undertaking involving five space agencies, where each partner contributes in a manner comparable to its technical expertise and development.[5] From a technical point of view, the ISS is a research laboratory orbiting the earth, taking advantage of the micro-gravity environment to conduct experiments.[6] Although space stations were already an established practice with the first space station being launched by the Soviet Union in 1971,[7] the previous space stations were both operationally and legally simpler. The ISS is particular since it consists of different flight elements (modules), registered by different Partners, and thus it demands the cooperation of several states.

The development of the ISS began after the invitation of the United States president Reagan to other countries to participate in the establishment of a permanently inhabited civil space station in Earth's orbit.[8] Many years of negotiations followed which led to the signing of the Intergovernmental Agreement (IGA) of 1988.[9] After the fall of the Soviet Union, Russia was invited in the cooperation which led to the amendment of the original Agreement in 1998. The IGA of 1998[10] has been accepted by fifteen partner States as well as by five Cooperating Agencies.[11] It is of course consistent with the outer space treaties, but it deals with a number of issues that were not previously addressed in space law. It represents the legal core of the ISS cooperation, proving to be a successful legal platform with potential to be used as a model for future space cooperation.

[5]Carla Sharpe, Fabio Tronchetti, Legal aspects of public manned spaceflight and space station operations in Frans von der Dunk, Fabio Tronchetti (eds) Handbook of Space Law (Edgar Elgar, 2015) 619.

[6]Ibid 620.

[7]Nola Taylor Redd, 'Salyut 1: The First Space Station' (Space, 26 July 2012). https://www.space.com/16773-first-space-station-salyut-1.html. Last accessed 15 September 2020.

[8]Rochus Moenter, 'The International Space Station: Legal Framework and Current Status' 64 (1999) 4 Journal of Air Law and Commerce 1033.

[9]Agreement among the United States of America, governments of Member States of the European Space Agency, the government of Japan, and the government of Canada on Cooperation in the Detailed Design, Development, Operation, and Utilization of the permanently Manned Civil Space Station, 29 September 1988 (1988 IGA).

[10]Agreement Among the Government of Canada, Governments of Member States of the European Space Agency, the Government of Japan, the Government of the Russian Federation, and the Government of the United States of America concerning Cooperation on the Civil International Space Station, 29 January 1998 (entered into force 27 March 2001) (1998 IGA or IGA).

[11]Sharpe, Tronchetti (n 5) 631.

6.2 Establishing Extra-Terrestrial Jurisdiction

It seems clear that humanity has reached the reality of persons living for extended periods of time in outer space outposts. An establishment on Mars with potentially permanent human settlements, would require solutions to legal issues long predicted by Justice Brenan, according to who: 'we'll soon have to grapple with the question... what law should govern within space societies themselves and among space inhabitants who will people space communities."[12] One of the most important issues to address is the question of which jurisdiction would apply on a Mars settlement.

The IGA offers useful provisions that could be used as a basis for a potential Mars Agreement. Article 5 of the IGA provides the general jurisdictional framework for the ISS. In doing so, the first paragraph of Article 5 relies on the jurisdictional principles provided by the Outer Space Treaty and Registration Convention. As it is now custom in outer space law, a nation which registers a space object pursuant to Article II of the Registration Convention, can then, pursuant to Article VIII of the Outer Space Treaty, retain jurisdiction over such object or, in the case of the ISS, flight element.[13] Paragraph 1 makes reference to such principles confirming a territorial principle of jurisdiction, as the registered flight elements form a sort of 'quasi-territory' of the state that registered them.

In addition to this, Article 5 provides a jurisdictional foundation based on the nationality principle of jurisdiction. The second paragraph of the Article dictates that 'each Partner shall retain jurisdiction and control over the elements it registers in accordance with paragraph 1 above and over personnel in or on the Space Station who are its nationals'. A partner may thus exercise jurisdiction over acts occurring on its flight elements, but also over acts committed by personnel who are its nationals.

However, the most important provision in the IGA regarding jurisdiction on board the ISS is Article 22. The IGA is the first and only international legal text dealing with criminal jurisdiction in outer space, providing a set of rules that form a valuable precedent for any outer space establishment to come. Nevertheless, this cannot be said for the first iteration of the IGA. Indeed, the 1988 IGA has been criticized as a negative precedent for what concerns criminal jurisdiction.[14] The US was awarded an extra- ordinary jurisdictional grant, which practically granted the right to exercise jurisdiction for any given offence.[15] The following review concerns Article 22 of the 1998 IGA, which is generally regarded as a positive and valuable precedent.[16]

Once Russia became involved in the ISS cooperation, the US ceased to be the clearly dominant partner in terms of expertise and equipment provided.[17] The change

[12]Hans P Sinha, 'Criminal Jurisdiction on the International Space Station' (2004) 30 Journal of Space Law 85, 86.

[13]Ibid 107.

[14]See Stacy J Ratner, 'Establishing the Extraterrestrial: Criminal Jurisdiction and the International Space Station' (1999) 22 B C Int'l & Comp L Rev 323, 331.

[15]Ibid 332–333.

[16]Ibid 339.

[17]Ibid 334.

in Article 22 reflects the changed power structure, as all partners are awarded equal jurisdictional rights. The first paragraph of Article 22 provides the criminal jurisdiction basis, which is based on the nationality principle. Indeed, paragraph 1 dictates that the Partners 'may exercise criminal jurisdiction over personnel in or on any flight element who are their respective nationals'. This provision is interesting since the Partners decided not to include the principle of territoriality as a general basis of criminal jurisdiction. Establishments in outer space, and specifically the ISS, differ from their earthly equivalents since the concept of territorial limits is almost inexistent. An astronaut on the ISS floats across the different modules registered by different Partners and hence, it was decided to temper Article's 5 grant of general jurisdiction by removing territoriality as a principle for what concerns criminal jurisdiction.

The second paragraph on the other hand, broadens the possible criminal jurisdiction on board the space station to include both the passive nationality and territoriality principles. These provisions regulate cases where there is concurrent jurisdiction. Hence, when a criminal act includes an offender of one Partner state, that Partner has primary criminal jurisdiction over its national.[18] Nevertheless, that Partner at the request of any affected Partner State, meaning a Partner State whose national was a victim of the alleged criminal conduct, or on whose flight element the alleged crime occurred, must consult with the affected Partner concerning their respective prosecutorial interests. Following the consultations, the affected Partner may exercise criminal jurisdiction over the alleged perpetrator provided that within 90 days from such consultation, or another timeframe mutually agreed, the Partner State whose national committed the offense, either agrees to the affected Partner exercising its jurisdiction, or fails to provide assurances that it will submit the case to its competent authorities for the purpose of prosecution.[19]

Even if the personnel on the ISS are highly trained individuals, the drafters were concerned about crimes occurring on the space station, and thus, established a criminal jurisdiction regime. A settlement on Mars would most likely include not only trained personnel granted a specific technical function, but persons such as the families of the crew members, tourists or even persons moving to Mars in a plan of colonization. Providing a regime regulating criminal offences will be as important as it is down on Earth. In establishing future frameworks for criminal jurisdiction, Article 22 will not go unnoticed as it brings together several different concepts in one well-reasoned and well-balanced policy.[20]

6.3 Establishing Liability

Another aspect for which the ISS IGA would provide a useful precedent is liability. The IGA provides a rather special regime dedicating two articles to liability. Firstly,

[18]IGA (n 10) art 22 para 1.

[19]Ibid para 2.

[20]Sinha (n 12) 118.

it is made clear that the liability solutions within the ISS project do not derogate from the existing framework dictated by the Liability Convention.[21] Indeed, Article 17 provides that if the ISS were to cause damage to a third state compensable under the Liability Convention, the Partner States, and specifically the relevant launching states, would be held liable.[22] The same applies if the damage was caused to a space object of an ISS Partner, as long as such object is not part of the ISS cooperation. In this case, the ISS partner state would qualify as a third state.[23]

On the other hand, the provisions regarding liability between the partnership are laid out in Article 16 of the Agreement. The IGA provided for a regime which applies to the ISS activities as a '*lex specialis*' in the context of the '*lex generalis*' that is the Liability Convention.[24] More specifically, this provision applies to the 'Protected Space Operations' which are subsequently defined in the Article, and include all activities relating to the construction and operation of the ISS, whether in orbit, in transit or on Earth.[25] For all these activities, Article 16 anticipates a cross-waiver of liability. This clause foresees that each partner state and its cooperating agency or their related entities, shall wave the presentation of a compensation claim resulting from the provisions of the Liability Convention, against another partner state, its cooperating agency or related entities, for damage sustained from the Protected Space Operations.[26]

The aim of the cross-waiver was to encourage participation in the exploration, exploitation and use of outer space through the ISS. It broadly includes all the activities related to the ISS, with the only exception of cases concerning willful misconduct, bodily injury and intellectual property rights.[27] It is an example of how states parties to the Liability Convention can exercise their sovereign rights to conclude agreements as between themselves, applying different liability rules.[28] Establishing settlements on Mars would benefit greatly from a similar provision, since the eventual partners would not risk harming the cooperation through liability claims between themselves, on the contrary they would rather enhance it by waving liability claims.

---

[21] For the Liability Convention regime see Liability Convention (n 4) art II, III, IV V.

[22] Sharpe, Tronchetti (n 5) 637.

[23] Ibid.

[24] Ibid 638.

[25] IGA (n 10) art 16 para 3 (f); A Farand, Jurisdiction and Liability Issues in Carrying out Commercial Activities in the International Space Station (ISS) Programme in F. G. Von Der Dunk, M. M. T.A. Brus (eds) The International Space Station: Commercial Utilization from a European Legal Prespective (Martinus Nijhoff Publishers 2006) 92.

[26] Ibid Farand.

[27] IGA (n 10) art 16 para 2 (c).

[28] Liability Convention (n 4) art XXIII.

6.4 Intellectual Property Rights

The IGA was specifically designed to enhance the scientific, industrial and commercial use of the ISS.[29] The promotion of industrial and commercial activities however, requires a regime dealing with Intellectual Property Rights (IPR). Since the outer space treaties did not contain specific binding obligations between signatory nations as to outer space intellectual property and data rights, the ISS partners faced the need to put in place an adequate legal and commercial environment.[30]

When ratifying the Intergovernmental Agreement, the partners ensured intellectual property law would apply onboard the ISS modules. Article 21 provides a specific IPR framework reflecting the increased awareness of the importance of IPR in the stimulation and development of technical innovations in space.[31] After providing the definition of IPR for the purposes of the ISS, the second paragraph of Article 21 states that 'for purposes of intellectual property, an activity occurring in or on a Space Station flight element shall be deemed to have occurred only in the territory of the Partner State of that element's registry'. This provision expressly confirms the jurisdiction of the state of registration for what concerns IPR, in the same way that Article 5 provides for general jurisdiction.

The Partners had envisaged the role of the ISS in the development of future products such as products in the realm of medicine, pharmaceuticals and new materials development.[32] This is the reason why Article 21 specifically makes mention to patents and their applicable regime, as they were considered important for the commercial development of the ISS.[33] Patent law is though complicated since the ISS Partners have different standards for granting patents and recognizing the patent holder, and a unified international approach would be difficult.[34] Given the unharmonized status both globally and regionally, it was easier to extended the partners' domestic IPR laws as to apply onboard the ISS on a quasi-territorial basis, instead of establishing a global and univocal legal framework. Being the ISS composed by a mix of flight elements registered by different Partner States, different patent laws apply to each module of the space station.

Article 21 further offers provisions for cases 'concurrent' cases, where it would not be clear which Partner's law would apply. Firstly, the third paragraph provides for cases where an invention is made on an ISS module by a person who is not its national. In those cases, a Partner state shall not apply its laws regarding secrecy of

---

[29] IGA (n 10) art 1 para 1.

[30] Lee Ann W Lockridge, 'Intellectual Property in Outer Space: International Law, National Jurisdiction, and Exclusive Rights in Geospatial Data and Databases' (2006) 32 Journal of Space Law 319, 331.

[31] Rosario Avveduto, 'Past, Present, and Future of Intellectual Property in Space: Old Answers to New Questions' (2019) 29 Washington International Law Journal 203, 226.

[32] Sharpe, Tronchetti (n5) 635.

[33] IGA (n 10) art 21 para 3.

[34] Alexandra M Davidson, 'To Explore Outer Space: The Intellectual Property Frontier for Patents' (2019) 47 Hofstra L Rev 889, 891.

inventions so as to prevent 'the filing of a patent application ... in any other Partner State that provides for the protection of the secrecy of patent applications containing information that is classified or otherwise protected for national security purposes'.[35] For instance, if an ESA astronaut makes an invention on a US registered module, ESA holds the right to choose the place where to file the patent without consideration of the US Inventions Secrecy Act.[36] The last paragraph further stipulates that the temporary presence in a different jurisdiction on the ISS or in transit to the ISS, cannot be the basis for patent infringement claims in the state where the transition occurs.[37] Article 21 follows the principle of temporary presence, pursuing the freedom of transfer of technology that could otherwise infringe intellectual property laws.[38]

Another aspect for which the IGA provides a solution is IPR in relation to ESA. ESA finds itself in a particular situation regarding IPR. Unlike the other states which are represented each by a single 'Cooperating Agency', the European states that participate in the ISS and form collectively the 'European Partner', are represented by ESA.[39] The flight elements of the European Partner are registered under ESA, which can also act as the state of registry.[40] The regulation of IPR relating to activities onboard the European Partner's flight elements hence, required a specific solution since ESA has no jurisdiction by itself. The solution chosen by the partners was to consider an activity occurring on an ESA module, as to have taken place in all the European Partner states simultaneously.[41] Consequently, the different national laws of the European Partner states could potentially apply for activities occurring on the European ISS modules, as each partner individually determines the conditions required for intellectual property protection in their jurisdiction.

The Intergovernmental Agreement is the first and only international agreement dealing specifically with IPR in space. The outer space treaties, or even the most important treaties regarding intellectual property protection, such as the Trade-Related Aspects of Intellectual Property Rights (TRIPS),[42] the Paris Convention,[43] or the Berne Convention,[44] have not expressly considered the issue of IPR in outer space.[45] Being signed by the major space powers, the Agreement forms an important

---

[35]IGA (n 10) art 21 para 3.

[36]US Invention Secrecy Act, 35USC s 184.

[37]IGA (n 10) art 21 para 6.

[38]Avveduto (n 31) 228.

[39]IGA (n 10) preamble.

[40]Ibid art 5 para 1.

[41]A M Balsano, J Wheeler, The IGA and ESA: Protecting Intellectual Property Rights in the Context of ISS Activities in von der Dunk, Brus (n 26) 68.

[42]Agreement on Trade-Related Aspects of Intellectual Property Rights (enacted 15 April 1994) 1869 UNTS 183 (TRIPS).

[43]Paris Convention for the Protection of Industrial Property (enacted 20 March 1883) as last revised on 14 July 1967, 828 UNTS 305 (Paris Convention).

[44]Berne Convention for the Protection of Literary and Artistic Works (enacted 9 September 1886) as revised on 24 July 1971, 1161 UNTS 3 (Berne Convention).

[45]Juan Felipe Jimenez, 'Patents in Outer Space: An Approach to the Legal Framework of Future Inventions' (2016) 98 J Pat & Trademark Off Soc'y 447, 456.

precedent for establishing the rules of IPR in space. Hence, when drafting an agreement to regulate human presence on Mars, it is certain that the provisions offered by the IGA will not go unnoticed.

6.5 Commercial Utilization—Space Tourism

Ever since outer space was reached by human-made satellites, it was found to be abundant of opportunities for commercialization. The ISS Partners realized the benefits the commercial exploitation of the space station would bring to the development of the project and included the enhancement of the commercial use of the ISS among the objectives of the cooperation.[46] The ISS legal framework represents a positive example of how an international project can exploit commercial participation in order to fund the operation of the programme, or promote the development of new technologies.

One of the most interesting commercial uses of the ISS has been undoubtably space tourism. In a broad sense, space tourism can be defined as 'any commercial activity offering customers direct or indirect experience with space travel'.[47] A space tourist can consequently be defined as 'someone who tours or travels into, to, or through space or to a celestial body for pleasure and/or recreation'.[48] However, the term 'space tourist' has not been defined from a legal point of view.[49] The outer space treaties were drafted with the interest of 'astronauts' or 'personnel', terms though that have not been defined either. Apart from some definitions and opinions offered by scholars, space tourism is a field that lacks legal certainty.[50]

Nevertheless, as it has been proven on the ISS, space tourism has significant potential advantages such as reducing operational costs for the Partners. In 2001, Dennis Tito, a private individual, travelled on board the Russian Soyuz capsule for a six-day stay on the ISS, becoming the first 'space tourist' on the space station,

---

[46]IGA (n 10) art 1 para 1.

[47]Stephan Hobe, Jurgen Cloppenburg, 'Towards a New Aerospace Convention? Selected Legal Issues of "Space Tourism"' (2004) 47 Proceedings of the Colloquium on the Law of Outer Space 377.

[48]Zeldine Niamh O'Brien, 'Liability for Injury, Loss or Damage to the Space Tourist' (2004) 47 Proceedings of the Colloquium on the Law of Outer Space 386.

[49]A Ferreira-Snyman, 'Legal Challenges Relating to the Commercial Use of Outer Space, with Specific Reference to Space Tourism' ' (2014) 17 Potchefstroom Elec LJ 1, 18.

[50]It has been suggested by some that space tourists should undergo some sort of training in order to ensure they receive the humanitarian protection offered by the Rescue Agreement; see Steven Freeland, 'Fly Me to the Moon: How Will International Law Cope with Commercial Space Tourism' (2010) 11 MELB J INT'l L 90, 98; see Tanja Masson-Zwaan, Steven Freeland, 'Between heaven and earth: The Legal Challenges of Human Space Travel' (2010) 66 Acta Astronautica 1597, 1604; others suggest that space tourists cannot be considered as personnel of the spacecraft since they do not perform functions relating to the operating of the space vehicle during their relatively short period in outer space; see Ibid, 20.

while marking the first time an ISS mission was commercially funded.[51] Since then, seven tourists have visited the ISS, all through commercial opportunities offered by the Russian Space Agency. Allowing private individuals to visit the ISS, has been a proven practice of the Russian Partner in order to support the funding of their part of the ISS operational costs.

Allowing though private individuals on board the ISS, raised several questions among the Partners. According to Article 5 of the IGA, Russia had the sovereign discretion to allow Tito on board its module.[52] Furthermore, According to Article 9 the partners have 'the right to barter or sell any portion of their respective allocations'.[53] Nevertheless, the Partners initially opposed to the participation of Tito, arguing that the presence of an 'amateur' on the ISS would endanger the space station and the permanent crew.[54]

Moreover, the ISS legal framework does not include any arrangement in the context of personal liability. Hitherto all persons who had entered into outer space were highly trained individuals, normally employees of governmental agencies with professional purposes.[55] This meant that the personnel on the ISS, being professionals involved in the 'Protected space Operations', enjoyed the waiver of liability, as long as the damage caused did not involve gross negligence or disregard of orders.[56] Consequently, there were concerns for a situation where Tito could cause damage to the space station or to experiments, without the possibility to hold Russia liable for any such damage.[57] The only possibility foreseen by the Intergovernmental Agreement was Partner States bringing in a further state, or private entity under their jurisdiction, following the consultation and a priori consensus of the other partners for doing so.[58]

The complication resulting from the potential damage caused by Tito to the ISS were solved with an ad hoc agreement between the relevant partners. In the end Tito was allowed to the ISS as long as his activities were limited to the Russian module, while Russia took out liability insurance.[59] However, as the Partners realized that bringing tourist to the ISS, as long as accordingly regulated, would benefit the commercialization of the space station, which, in its turn, would help to alleviate

---

[51] Edith Walter, The Privatization and Commercialization of Outer Space in Christian Brunner, Alexander Soucek (eds) Outer Space in Society, Politics and Law (Springer-Verlag/Wien 2011) 500; even if the transportation to the ISS was carried out by the Soyuz (governmental spacecraft) and the destination was an intergovernmental space station, the motivation and the money for the flight were of private commercial character; see Frans von der Dunk, Legal Aspects of Private Manned Spaceflight in von der Dunk, Tronchetti (n 5) 697.

[52] IGA (n 10) art 5 para 2.

[53] Ibid art 9 para 2.

[54] Steven Freeland, 'Up, up and … Back: The Emergence of Space Tourism and Its Impact on the International Law of Outer Space' (2005) 6 CHI J INT'l L 1, 3.

[55] Frans von der Dunk (n 51) 700.

[56] Ibid.

[57] Ibid.

[58] IGA (n 10) art 9 para 3 (a).

[59] See Freeland (n 54) 2.

governmental budget problems, the attitude changed giving rise to a comprehensive and formalized solution.[60]

In early 2002, the Cooperating Agencies of the partners reached the Multilateral Crew Operations Panel Agreement ('MCOP Agreement')[61] as to who was allowed on the ISS. The MCOP Agreement provides general guidelines for selection, assignment and training of the crew and it also defines the criteria regarding the certification of crew flight readiness.[62] Most importantly though, it clearly distinguishes between different categories of crew members allowed on the ISS.

The Agreement represents a first in space law, as for the first time the term 'astronaut' was defined in an international legal text.[63] Moreover, the MCOP Agreement added the category of 'spaceflight participants' which are defined as individuals sponsored by a Partner.[64] The Agreement further distinguishes crew members in 'expedition crewmembers', who form the main crew of the ISS and are responsible for 'implementing the planned activities for an increment',[65] and 'visiting crewmembers', who are not part of an expedition and may either be professional astronauts (or cosmonauts) or spaceflight participants.[66] The latter may include scientists and even space tourists with specific functions.

Even if the aforementioned provisions have a limited scope, von der Dunk believes that the distinctions between crew members offered by the MCOT Agreement may constitute a 'trendsetting, if not an industry standard'.[67] Other scholars also believe that the legal documents relating to the ISS may serve as examples towards the clarification of definitions regarding astronauts and the legal status of different participants in commercial spaceflight.[68] A Mars establishment and any outpost in outer space is expected to attract commercial activities and consequently tourists. Given the development of technology and the primary role of the private sector, space tourism is one of the most important markets for generating profit from outer space. Legal certainty in the sector is a fundamental requirement for its development. A new Agreement regarding the exploitation on Mars would need to create such legal certainty by offering definitions and applicable rules. In doing so, the ISS is undoubtably a handy precedent in a sphere that apart the MCOT Agreement there is a complete lack of definition.

---

[60] Von der Dunk (n 51) 701.

[61] The Multilateral Crew Operations Panel's Principles Regarding Processes and Criteria for Selection, Assignment, Training and Certification of ISS (Expedition and Visiting) Crewmembers (2001) (MCOP Agreement) visible at (Space Ref 2002) www.spaceref.com/news/viewsr.html?pid=4578, last accessed 15 September 2020.

[62] Stephan Hobe, 'Legal Aspects of Space Tourism' (2007) Nebraska Law Review 439, 458.

[63] MCOP Agreement (n 61) para III.

[64] Ibid.

[65] Ibid.

[66] Ibid.

[67] Frans G von der Dunk, 'A Sleeping Beauty Awakens: The 1968 Rescue Agreement after Forty Years' (2008) 34 Journal of Space Law 411, 433.

[68] Hobe (n 62) 457.

6.6 Conclusion

The ISS project has achieved a high level of structured international cooperation between the Partner States. The IGA and the rest of the multilateral and bilateral legal instruments, have developed an elaborate framework that has contributed in making the ISS a successful example of large-scale international cooperation in outer space.[69] Therefore, an outer space exploration agreement would find a successful legal platform in the IGA, a platform that has shown that governments and private entities can collaborate in technological, financial, political and legal levels to produce successful projects that provide for the benefit of all with little dispute and operational difficulty.[70]

The presence of mankind on another celestial body will raise many legal questions. Many of such issues, such as the exploitation natural space resources, or the legal status of permanent Mars stations, would require the development of new legal solutions. The ISS framework however, offers a way to approach some of these issues, providing unique and proven solutions in how to manage international cooperation with respect to long-duration outer space manned missions.

Alexandros Eleftherios Farsaris graduated from the University of Macerata, Italy with a thesis analysing the International Space Station's legal framework. Currently he is enrolled in the Master of Space, Communications and Media Law at the University of Luxembourg.

[69] Sharpe, Tronchetti (n 5) 659.

[70] See E González Ferreiro, A Azcárraga, 'Orbital Space Ports: Their Operating Procedures and Legal Frame' Proceedings of the Fiftieth Colloquium on the Law of Outer Space (2008) 160.

Chapter 7
Building a New Legal Model for Settlements on Mars

Ivan Fino

Abstract The quest for "infinity" of humans needs to keep climate change in mind. The current legal framework may discourage the development of the Mars settlements. If a company develops a facility, there may be a contrast with the freedom of access of Article 1, Outer Space Treaty. When a chattel is affixed to the land, it would become common property. Which regime could ensure legal certainty on Mars? An International Legal Trust System (ILTS) should be set up, which would be supervised by the United Nations Office for Outer Space Affairs (UNOOSA), as the main trustee. Specifically, when a company is interested making any operation, it will send a work plan to the UNOOSA. The plan of work shall be consistent with the legal parameters of economic and environmental sustainability; in case of lucrative sites, the company shall provide benefit-sharing. If the plan is approved, the country in which the company has been registered assumes the role of co-trustee, investigating if the national company follows the work plan. When the applicant for the space activity is a nation, the trustee would be the UN. Finally, the ILTS would grant property rights, while the income of benefit-sharing could finance global goals.

7.1 Introduction

Humans must look to Mars and outer space for several reasons. It should be considered that along the Earth's history there have been severe extinctions caused by natural events, such as nearby Galactic Gamma-Ray Bursts[1] or asteroid collisions.[2] Nevertheless, humankind's survival is already in peril because of its "qualitative"

[1]Piran T, Jimenez R (2014) Possible Role of Gamma Ray Bursts on Life Extinction in the Universe. Phys. Rev. Lett. 231102:1–6.

[2]Shen J et al. (2019) Evidence for a prolonged Permian–Triassic extinction interval from global marine mercury records. Nat Commun 10(1563):1–9.

I. Fino (✉)
Faculty of Law, University of Turin, Turin, Italy
e-mail: ivan.fino@edu.unito.it

A. Froehlich (ed.), *Assessing a Mars Agreement Including Human Settlements*, Studies in Space Policy 30, https://doi.org/10.1007/978-3-030-65013-1_7

alteration of the environment and ecosystems.[3] If climate change continues to get worse, Earth could become inhospitable and outer space may become the only place to preserve humankind, leading to the new epoch of "*Cosmigrocene*".[4] Bearing this in mind, Mars settlements may be the only way to preserve human race. It is time to look to Mars also to better human existence on Earth. Space mining may help to support and diversify the growing demand of materials critical to Earths' needs, such as platinum group metals and rare Earth elements. Moreover, the offshoring of heavy industry could help to preserve the environment. These and other activities of the new space economy may raise several legal and policy issues because the current legal framework does not furnish an effective space resource allocation. Furthermore, it should be considered that the legal status of celestial bodies' resources was created in the sixties, when space activities were at a pioneering stage. The first aim of the international community then, was to avoid the militarisation of outer space.[5]

The aim of this article is to analyse the actual legal status of outer space (first part), to focus on the legal issues that could arise in the future (second part), to explain why the Moon Agreement has not achieved wide acceptance in the international community (third part), and finally propose a new international legal regime for outer space to both encourage space economy and international cooperation among developed and developing nations (fourth part).

7.2 The Current Legal Status of Outer Space

The general legal status of the cosmic domain is provided by the first two articles of the 1967 Outer Space Treaty.[6] While Article I provides the four main freedoms regarding outer space (freedom of use, exploration, access, and scientific investigation), Article II states the main limits relating to human activities in outer space, namely the non-appropriation principle by claim of sovereignty, by means of use or occupation or by any other means. The joint interpretation of the two key concepts "*freedom of use*" of Article 1 and "*prohibition of appropriation*" of Article 2, allows us to determine which space activities are permitted and which legal issues may arise. Since the Outer Space Treaty does not specify any further details of the words *use*, it should be interpreted in accordance with the Vienna Convention of the Law of the

[3] Arias-Maldonado M (2015) Environment and Society. Socionatural Relations in the Anthropocene. Springer, Cham (Switzerland).

[4] Cosmigrocene is a neologism that derive from the Latin "*Cosmos*" (space), "*Migros*" (migration), and from the Greek "*Kainos*" (recent).

[5] Tronchetti F (2007) The Non-Appropriation Principle Under Attack: Using Article II of the Outer Space Treaty in Its Defence. The 68th International Astronautical Congress Symposium. https://iislweb.org/docs/Diederiks2007.pdf. Accessed 30 July 2020.

[6] United Nations Office for Outer Space Affairs (UNOOSA): Treaty on principles governing the activities of states in the exploration and use of outer space, including the moon and other celestial bodies ("Outer Space Treaty") (1966). https://www.unoosa.org/oosa/en/ourwork/spacelaw/treaties/outerspacetreaty.html. Accessed 31 July 2020.

Treaties of 1969.[7] The first interpretation method featured in the provision points to the ordinary meaning of the term, in its context and considering its object and purpose. *Use* could be addressed as taking from a supply, employing for a purpose, converting a thing for a goal,[8] or taking the control over a property oriented to get profit or other benefits.[9] It should also be evidenced that the purpose of the Outer Space Treaty is to promote the exploration and use of outer space and not to restrict them, as emerges in its preamble. Hence, the meaning of what is allowed should be deduced from what is not forbidden[10] because "where activities are not expressly prohibited, ambiguities should be construed in a permissive rather than restrictive way to avoid impeding the development of the uses of outer space".[11] Given that the non-appropriation rule of Article 2 of the Outer Space Treaty, does not mention outer space resources, *the appropriation of them should be admitted*. Finally, the freedom to use and the appropriation-prohibition could be converged in one system founded on a non-exclusive use, in which one or more entities can extract resources, but cannot claim any ownership over the site of extraction. The extraction of materials would be allowed, opening the doors to the commercial uses of outer space, without any expressed limit for anyone to carry out the exploitation of rare earth elements and ice. Furthermore, the aforementioned interpretation may be considered as customary law,[12] binding both for signatories and non-signatory States. As stated by the ICJ in the Nicaragua case[13] and in the Continental Shelf case[14] customary law consists of two elements, namely the *general practice* and *opinio juris* of States. From the first Apollo Mission of 1969, the United States has considered as government property, only the collected lunar materials, but not the site of extraction.[15] In the following years, the USSR as well as other countries followed the same practice in carrying out sample return missions to Moon, Mars and asteroids. No opposition to this practice

[7]United Nations (UN): Vienna Convention on the Law of Treaties (1969). https://treaties.un.org/Pages/ViewDetailsIII.aspx?src=TREATY&mtdsg_no=XXIII-1&chapter=23&Temp=mtdsg3&clang=_en. Accessed 30 August 2020.

[8]Simpson J A, Weiner E S C (1989) The Oxford English Dictionary, XIX volume. Clarendon Press, Oxford (U.K.).

[9]Garner B A (2009) Black's Law Dictionary, West, Eagan (U.SA.).

[10]Stephens D (2017) Recent Developments in Space Law: Opportunities and Challenges. Springer, Singapore.

[11]Lintner A (2016) Extraterrestrial Extraction: The International Implications of the Space Resource Exploration and Utilization Act of 2015. Fletcher Forum of World Affairs. https://static1.squarespace.com/static/579fc2ad725e253a86230610/t/57ec6ac65016e1636a21e331/1475111622859/FletcherForum_Sum16_40-2_139-157_LINTNER.pdf. Accessed 20 August 2020.

[12]Boaz D, HUDGINS E L (2002) SPACE: THE FREEMARKET FRONTIER. Cato Institute, Washington (U.S.A.).

[13]International Court of Justice(ICJ): Reports of Judgments, Advisory Opinions and Orders. Case Concerning Military and Paramilitary Activities in and Against Nicaragua (1986). https://www.icj-cij.org/files/case-related/70/070-19860627-JUD-01-00-EN.pdf. Accessed 20 August 2020.

[14]International Court of Justice(ICJ): Reports of Judgments, Advisory Opinions and Orders. Case Concerning the Continental Shelf (1985). https://www.icj-cij.org/files/case-related/68/068-19850603-JUD-01-00-EN.pdf. Accessed 28 August 2020.

[15]Kleiman M J (2013) The little book of space law. American Bar Association, Chicago (U.S.A.).

has been offered by any other nation.[16] Moreover, regarding the second element of customary law, namely *opinio juris*, some nations have enacted domestic legislation on the commercial exploitation of outer space. The title IV of the US Commercial Space Launch Competitiveness Act of 2015 declares that "the United States does not, by enactment of this Act, assert sovereignty, exclusive rights or ownership of any celestial body".[17] The same act grants U.S. citizens the right to possess, own, use and sell "space resources obtained." In 2017, Luxembourg has legally recognised that "space resources are capable of being appropriated".[18] Other space faring nations are considering similar laws, among them Russia, UAE, Japan, China, and Australia.[19]

7.3 The Inadequacy of the Actual Legal Regime

Having explained the actual legal regime of outer space, there may be an array of issues that could hamper future space activities, some of which will be further examined. What if a company, such as the Planetoid Mines or the Rio Tinto Group would engage into extract water or other materials? What if another company, such as Space X, adapts a portion of a celestial body for any other need? No entity has the right to exclude any others from space resource extraction or from use of the land. Moreover, there is no rule currently deciding who takes precedence in the use. This situation could lead to several possible outcomes, such as complex litigations or the "tragedy of the common space" theorised by the economist Garrett Hardin.[20] Another possible legal issue would be the complete disruption of a celestial body from the total use of its resources. In this situation, the use of the celestial body resources would be equal to its appropriation, prohibited by article II. Moreover, a chattel (in which ownership is retained) becomes a fixture when it is affixed to land, becoming common property.

Therefore, these possible scenarios could hinder the development of the space economy itself. The legal and political uncertainties could stop companies from investing in space mineral exploration and development. As much as technological development will allow the exploitation of outer space resources, the economic advantages could be incentive enough for nations with space capabilities to push for a break of the Outer Space Treaty. Nations may try to ignore or to reinterpret the

---

[16]See Hudgins, supra note 13.

[17]U.S. Commercial Space Launch Competitiveness and Entrepreneurship Act, Pub. L. No. 114-9 (2015), https://www.congress.gov/114/plaws/publ90/PLAW-114publ90.pdf. Accessed 25 August 2020.

[18]Loi du 20 juillet 2017 sur l'exploration et l'utilisation des ressources de l'espace. http://legilux.public.lu/eli/etat/leg/loi/2017/07/20/a674/jo#:~:text=Aucune%20personne%20ne%20peut%20explorer,apr%C3%A8s%20%E2%80%9Eles%20ministres%E2%80%9C. Accessed 25 August 2020.

[19]Pershing A D (2019) Interpreting the Outer Space Treaty's Non-Appropriation Principle: Customary International Law from 1967 to Today. Yale J. Int'l L. 44:149–178.

[20]Hardin G (1968) The Tragedy of the Commons. Science 162:1243–1248.

non-appropriation clause, leaving outer space without certain specific rules accepted by the international community, that could lead to chaos and anarchy.

7.4 The Missed Opportunity of the Moon Agreement

The 1979 Moon Agreement[21] attempted to establish preconditions for a regime of space exploitation. It has not been widely accepted by the international community because of several reasons. Firstly, developed nations may have seen the prohibition of ownership as a deterrent for private investment. It should be considered that during the end of the seventies the possibility of private space activities was more concrete rather than in the sixties, when the Outer Space Treaty was adopted. Another controversial point was how much benefit sharing companies would have to pay.[22] There are also practical legal inconsistencies that may arise from the application of both the Moon Agreement and the Outer Space Treaty. While the Moon Agreement grants the possibility to build facilities, stations and installations or other fixtures on celestial bodies, there could be a contrast with article 1 of the Outer Space Treaty that declares the freedom of access of all areas of outer space. In other words, a facility or a station could interfere with the right of access to all areas of outer space. This possible inconsistency could happen for example on Mars, due to the need to render the surface agriculturally productive or habitable. Finally, the Moon Agreement is not legally binding on any non-party State because it is not customary law.[23]

7.5 A Proposal for an International Legal Trust System

Since several legal and policy issues may arise from the actual legal framework, a new international legal regime for outer space shall:

(a) Provide for property rights or a lease allocation system, both incentivising investments in the space sector. The system would be supervised and led by the United Nations (UN) through the United Nations Office for Outer Space Affairs (UNOOSA).
(b) Establish the rule of law in outer space. A laissez faire system could turn into anarchy whereby countries and companies could race to grab as many resources as possible bringing considerable potential conflict.

---

[21] United Nations Office for Outer Space Affairs (UNOOSA): Agreement Governing the Activities of States on the Moon and Other Celestial Bodies ("Moon Agreement"), 1979. https://www.unoosa.org/oosa/en/ourwork/spacelaw/treaties/intromoon-agreement.html. Accessed 31 August 2020.

[22] Paragraph 5, Article 11, Moon Agreement: "States Parties...undertake to establish an international regime...to govern the exploitation of the natural resources of the moon...".

[23] Baca A K (1993) Property Rights in Outer Space. J. Air L. & Com. 58:1041–1085.

(c) Recognise outer space as *common heritage of mankind*, instead of *res communis*.[24]
(d) Provide a sustainable exploitation of celestial bodies, to avoid the uncontrolled production of space debris or to prevent the complete exhaustion of the celestial bodies' masses or their natural orbits.[25] The United Nations should manage the ordered and sustainable economic development in outer space for the present and future generations.
(e) Prevent the militarisation of outer space and favours the international collaboration, which are the same aims of the Outer Space Treaty' drafters.
(f) Consider the weak points of the Moon Agreement which led to nations' refusal to sign. Only a widely accepted agreement would have the power of law in the international context.

The abovementioned requirements could be met by establishing an international Legal Trust System (ILTS). A trust is an arrangement that assigns assets to one or more trustees that will manage them in the interest of one or more beneficiaries. The latter may include the trustee or the settlor.[26] Translated in the ILTS, mankind would assume the role of settlor and beneficiary of the outer space resources. The UNOOSA would act as main trustee of outer space resources and trading property rights and leases to companies and countries. The rights over the celestial bodies or over its resources would depend on the nature of the celestial body itself. For example, property rights are preferable to a lease over asteroids, as they could just disappear after the exploitation. Both leases and property rights can be provided over lands and mining sites on Mars. Leases or defeasible titles are preferable for some land mass on those celestial bodies which could hypothetically be used by humankind pending an Earth disaster. In the case of lucrative activities, such as mining, companies will choose whether to get the exclusive use over the resource through payment of the lease or through annual payment linked to net proceeds or to production charges.

7.6 The Functioning of the International Legal Trust System

When a company is interested in leasing or buying an outer space resource, before starting any operations, it must send a plan of work to the United Nations. The plan of work shall include all the details of the activity that would be carried out; it shall be consistent with pre-established parameters of sustainability and shall not interfere with other space activities. If the UN approves the company plan of work, the country

[24]There is a fundamental difference between the "res communis" concept and "Common Heritage of Mankind". The latter implies that if any resource was extracted from a common area, any economic benefits would be shared internationally (Joyner C (1986) Legal Implications of the Concept of the Common Heritage of Mankind. Int'l & Comp. L. Q. 35:190–199.

[25]See Lee, supra note 12.

[26]Martin E A (2006) Oxford Dictionary of Law. Oxford University Press, Oxford (U.K.).

of the company assumes the role of co-trustee for the specific resource. Thus, as a co-trustee, countries must investigate whether all activities of their national companies are consistent with the plan of work authorised by the UN. These supervisory duties would be added to the responsibility of nations for all space objects that are launched within their territory.[27] The UN, as main trustee, would oversee that countries are performing their duties. This model would be the *ordinary* one. There would be also an *extraordinary* model, in which the UN would be the only trustee. This model would be possible in two instances: when the country of the applicant for a private company is not technologically able to act as a trustee or when the applicant of the activity is a country itself. Furthermore, as stated previously, the beneficiaries of this trust are the countries of the world and their citizens; hence all mankind would take concrete profit from lease transactions and benefit sharing. The income from the sales, leases and benefit sharing can be distributed to mankind by financing international global goals, following a similar model of the 17 Sustainable Development Goals adopted by the United Nations in 2015, which addressed poverty, inequality, climate change, environmental degradation, and peace and justice. Finally, the International Legal Trust System would meet acceptance because every country would obtain benefit sharing to improve its living standard and space faring nations would rely on property rights.

7.7 How to Implement the International Legal Trust System

In order to enact the International Legal Trust System, the articles 1 and 2 of the Outer Space Treaty shall be amended by following the procedure provided by article 15 Outer Space Treaty.[28]

The modified version of the two articles is illustrated below:

Article I

The exploration and use of ***activities in*** *outer space, including the moon and other celestial bodies, shall be carried out for the benefit and in the interests of all countries, irrespective of their degree of economic or scientific development and shall be the province of all mankind.*

Outer space, including the moon and other celestial bodies, is the common heritage of mankind and its purpose is to improve and ensure the existence of humanity and world peace.

Subject to Article II, outer space, including the moon and other celestial bodies, shall be free for exploration and use by all States ***government and non-government entities****, without discrimination of any kind, on a basis of equality and in accordance with international law, and there shall be free access to all areas of celestial bodies.*

---

[27] Article 6, Outer Space Treaty.

[28] Article 15, Outer Space Treaty: "Any State Party to the Treaty may propose amendments to this Treaty…".

There shall be freedom of scientific investigation in outer space, including the moon and other celestial bodies, and States shall facilitate and encourage international cooperation in such investigation.

Article II

Outer space, including the moon and other celestial bodies, is not subject to ***unilateral*** *national appropriation by* ***governmental or non-governmental entities*** *claim of sovereignty, by means of use or occupation, or by any other means.*

To promote the international collaboration and to ensure the common heritage principle, property and exclusive rights on celestial bodies shall be admitted.

Moreover, operational details of the International Legal Trust System and the economic and environmental parameters of sustainability should be provided by a supplementary annex, as provided for the United Nations Convention on the Law of the Sea.[29]

7.8 The Advantages of the International Legal Trust System

The International Legal Trust System would outperform the actual legal framework for outer space, as well as past agreements for the "common space areas" (Antarctic Treaty, UNCLOS III and Moon Agreement). The advantages of an International Legal Trust System would be:

(a) The capability of efficiently organising the interactions among nations on one hand, and third parties or companies on the other hand. This makes the trust perform a property law-like function because it provides a system in place rather than just a contract law-like function. This can be done efficiently through the UN, which can use UNOOSA as its secretarial office for performing everyday functions.
(b) It would have an efficient and flexible organisational structure, compared to legal frameworks based on corporate law or liberal business corporation statues.
(c) It assigns property rights better than contract law.
(d) It allows to have the functioning of both contract law and property law.
(e) Finally, companies would deal directly with their government administration rather than with the UN. The procedure provided by the UNCLOS III binds companies to deal only with the UN or its authorities (International Seabed Authority) which would be more complex and costly.

[29] United Nations (UN): Agreement relating to the implementation of Part XI of the United Nations Convention on the Law of the Sea (UNCLOS III 1982). https://www.un.org/Depts/los/convention_agreements/texts/unclos/closindxAgree.htm. Accessed 27 August 2020.

Ivan Fino is a graduate from Roma Tor Vergata (Italy) in “Legal Security Sciences.” His dissertation was on Environmental Criminal Law. Currently he is working on a Master’s degree in Law at the University of Turin, Italy. His research thesis is about the application of a legal trust system for outer space.

Chapter 8
Factors Influencing the Future Martian Population

Satyam Tiwari

Abstract With the increasing understanding of the Martian Environment, Mars is being considered as a possible destination for human settlement. Several space agencies like NASA, SpaceX, ESA, UAE Space Agency, and CNSA are taking a new initiative to start projects for human settlements on Mars. Humanity's ambition for becoming an interplanetary species can become possible by making a leap to Mars. Settlement on Mars would not only offer an alternate shelter to humankind but also help in preserving the human species from a catastrophic event on Earth. With all characteristics that Mars shares with Earth, Mars seems to be the viable option to initiate interplanetary human settlement. In order to establish a self-sustaining settlement, humans will eventually start using Martian resources; but here a question arises: how many people can Mars sustain? This article addresses the factors that can affect the Martian population. A discussion of how the population for future Martian settlers depends on factors such as energy, area, technological advancement, political agendas, and engineering structured living quarters is introduced here. It is suggested that the Martian population also depends on technological advancement and political frameworks established by decision makers.

8.1 Introduction

Mars has always fascinated humans either about speculations of alien life on Mars or the future human settlement. Mars draws humanity's attention because Mars, unlike any other planet in the solar system, shares some standard features with Earth, including its axis tilt, seasons, surface features, day length, and the atmosphere.[1]

Mars, just like Earth, has an atmosphere, although it is much thinner and composed of different composition of gases. It generally consists of 95.32% of carbon dioxide,

[1] Amanda Rumble, 'What do Earth & Mars have in common' (*Sciencing*, 24 April 2017). www.sciencing.com/mars-earth-common-10034859.html. Accessed 10 July 2020.

S. Tiwari (✉)
Department of Aeronautical/Aerospace Engineering, Kurukshetra University, Haryana, India
e-mail: satyam.tiwari.aero@gmail.com

A. Froehlich (ed.), *Assessing a Mars Agreement Including Human Settlements*, Studies in Space Policy 30, https://doi.org/10.1007/978-3-030-65013-1_8

2.7% of nitrogen, 1.6% of argon, and trace levels of carbon monoxide, water, and oxygen.[2] With regard to mineral properties, Mars has silicon dioxide, iron, aluminium, calcium, potassium, and other valuable minerals that can act as source materials for human settlements on Mars. Additionally, just like Earth, Mars has northern and southern polar caps with a diameter of 1000 and 350 km across.[3] However, unlike Earth, these polar caps are composed of dry ice (carbon dioxide ice) and water ice. The northern cap alone is estimated to deposit 197,000 cubic miles of water, which could play an essential role in providing water for the Martian settlement.

The seasonal cycle on Mars supplies a continuous deposition of dry ice onto the polar caps, which could provide Martian settlers another tactical advantage to translate dry ice into oxygen and then eventually into water using ISRU (In-Situ Resource Utilization) implications, to support human life on Mars. As technology continues to advance, humans are moving one step closer to the practical, efficient utilization of ISRU. It may only be a matter of time before humans can utilize the resources available on Mars for sustainable living, scientific research, and commercial use.

Any Martian settlement would require a reliable technology and governance system to operate efficiently. After the few years of establishing the first Martian settlement, there may be a growth in the Martian population either naturally from births on Mars, or by continuous immigration from Earth, such as from space tourism or the space industry. A dynamic technological growth and a stable governance system could ensure a sustainable living to an ever-expanding Martian population. In this article, it is assumed that only humans with skilled training and appropriate educational backgrounds would be sent to Mars, so that any settlers are prepared for the technical challenges of the Martian environment.

This article evaluates some critical factors on which the Martian population could depend. At the same time, numerical approximation has been performed to give some insights about the amount of area and energy would be required by an individual Martian settler for sustainable living. An attempt to organize such factors in a hierarchical way is depicted in Fig. 8.1, where lower-level factors could have the least effect on the Martian population and higher/top-level factors could have the most immediate effect in controlling the Martian population.

8.2 ISRU Advancements

ISRU is the practice of collecting, storing, processing and translating one kind of resource into another. The technological advancement of ISRU plays a crucial role in determining the Martian population.

[2]Tim Sharp, 'Mars Atmosphere: Composition, Climate & Weather' (*Space.com,* 12 September 2017). www.space.com/16903-mars-atmosphere-climate-weather.html. Accessed 12 July 2020.

[3]Mira.org, 'Mars' (*MIRA*). www.mira.org/fts0/planets/097/text/txt002x.htm. Accessed 12 July 2020.

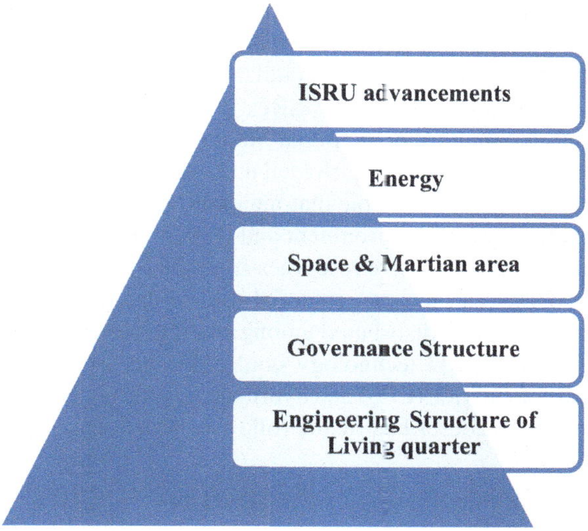

Fig. 8.1 A hierarchical overview of different factors affecting Martian population

The development of MOXIE (Mars Oxygen ISRU Equipment) in the *Perseverance Rover* of Mars 2020 mission is the first-ever inbuilt component to utilize ISRU for producing oxygen (O_2) from carbon dioxide (CO_2) in the Martian environment.[4] Such progress demonstrates that if science and engineering continue to advance, then it might be possible for humans to start initiating a permanent human settlement on Mars by the year 2040 or 2050.

Such technological advancement in ISRU would help humans to develop strategies for delivering all essential resources to Martian settlers sustainably. Some countries like the U.S and Luxembourg have already passed legislation to induce the commercialization of asteroids and Near-Earth Objects by utilizing ISRU. To effectively utilize the resources of the asteroids, the United States passed the *Space Resource and Exploration Act* or Space Act in 2015. The Space Act only allowed individuals and entities associated with the United States to commercialize the resources of asteroids using ISRU practices. The Space Act also does not directly address other states' rights over an asteroid, other than the United States. As a result, the Space Act, has raised concerns about whether or not U.S obligations under the Outer Space Treaty have been disregarded by enabling the commercialization of asteroid resources all alone.[5]

Luxembourg did not see the U.S Space Act as a solitary space act. It passed its act on *Exploration and Use of Space Resources* on 20 July 2017 for commercializing

---

[4]NASA, 'MOXIE for Scientists' (*NASA Science Mars 2020 Perseverance Rover).* https://mars.nasa.gov/mars2020/spacecraft/instruments/moxie/for-scientists/. Accessed 12 July 2020.

[5]Fabio Tronchetti, 'The Space Resource Exploration and Utilization Act: A move forward or step back?' (2015) 34 Space Policy. https://doi.org/10.1016/j.spacepol.2015.08.001. Accessed 18 July 2020.

Near Earth Objects using ISRU practices.[6] Unlike the U.S. Space Act, the Luxembourg Space Act provides every licensed individual and corporation under the E.U. the ability to start commercialization on Near Earth Objects.

Another factor under ISRU, which could affect the Martian population, is the development of international Martian legislation, under which the practices of ISRU would be performed, provided that every state has a right to settle and commercialize on Mars.[7] It is possible that international agreements might attempt to reduce stress on the Martian environment and thus limit the ability of settlements or corporations to extract and utilize resources from the Martian region. In this way, the application of ISRU can play an essential role in determining the amount of resources to be extracted and distributed among Martian settlers.

Such ISRU technology could also be used to produce methane (CH_4) using carbon dioxide (CO_2) and hydrogen (H_2) through the Sabatier reaction.[8] The Sabatier process is described by the following reaction:

$$CO_2 + 4H_2O \underset{\text{pressure}}{\overset{400\,^\circ C}{\rightarrow}} CH_4 + 2H_2O$$

Also, electrolyzing water ice or subsurface hydrated minerals could produce pure oxygen (O_2) and hydrogen (H_2) on Mars, which is described by the following reaction:

$$2H_2O + 4h\nu \rightarrow O_2 + 2H_2O$$

The products of these reactions could contribute to life-support systems and provide elements of bipropellant fuel for Martian return missions. A recent evaluation concluded that cryofreezing of CO_2 could be the most promising technology to convert it into CH_4 and O_2 efficiently. NASA's Johnson Space Center has undertaken the MARCO POLO (Mars Atmosphere and Regolith Collector/Processor for Lander Operation) project to design a CO_2 freezer for the Sabatier reactor.[9]

[6]ALLEN and OVERY, 'Luxembourg Space Resource Act: Paving the legal road to space' (*ALLEN and OVERY*, 28 September 2017). www.allenovery.com/en-gb/global/news-and-insights/publications/luxembourg-space-resources-act-paving-the-legal-road-to-space. Accessed 20 July 2020.

[7]Laurent Thailly, 'Luxembourg sets to become Europe's commercial space exploration hub with new Space law' (*Resource World, 2017*). www.resourceworld.com/luxembourg-set-become-europes-commercial-space-exploration-hub-new-spacelaw/#:~:text=Luxembourg%20set%20to%20become%20Europe's%20commercial%20space%20exploration%20hub%20with%20new%20Space%20Law,-Share%20Tweet%20Pin&text=Luxembourg%20is%20the%20first%20European,projects%20in%20the%20space%20sector. Accessed 25 July 2020.

[8]Master Class, 'What Is the Sabatier Reaction? Learn How Astronauts Create Water and Oxygen in Space' (*MasterClass*, 31 August 2019). www.masterclass.com/articles/what-is-the-sabatier-reaction. Accessed 25 July 2020.

[9]NASA TechPort, 'Mars atmosphere and Regolith Collector/Processor for Lander Ops (MARCO POLO) Atmospheric Processing Module (MARCO POLO APM)' (*NASA TechPort*). https://techport.nasa.gov/view/16846. Accessed 25 July 2020.

ISRU applications are not limited to water (H_2O) and oxygen (O_2) processing, but it can also be used to extract some buffer elements like (N_2) and argon (Ar) from the Martian atmosphere and regolith, to utilize it as a life support system. The discovery of 1100 ppm of nitrates in Martian regolith by *Curiosity's* S.A.M. (Sample Analysis at Mars) suite in 2015 suggests that these nitrates and additionally NH_4^+ in Martian soil could act as an active source for the production of N_2 for future Martian missions.[10]

Concluding the discussion, ISRU can help to form O_2 and CO_2 using reverse fuel cells and then into H_2O, which are the two vital components for enabling human settlement on Mars. ISRU can also help in extracting the N_2 as a by-product of the burial nitrates and ammonium minerals in the Martian regolith.[11]

Even with the constraints related to the Martian atmosphere, the application of ISRU could allow the cultivation of edibles on the Martian surface from plant-based sources, insect farming, and cellular agriculture under pressurized volumes of high LED lighting.[12] Ongoing research also indicates the possibility of increased efficiency by cultivating with soil-less systems on or under the Martian surface, provided that the nutrient cycle, perchlorate value, fertilizer amount, and texture of the Martian regolith is maintained. Such food sources would require the establishment of food facilities and bioreactors through locally available resources found on Mars.

8.3 Energy

Another factor that would determine the Martian population is the distribution of the energy by established power plants and reactors. Energy is going to be a vital requirement for a human settlement on Mars. Unlike Earth, Mars will probably be much more industrial because, on Mars, one cannot extract any useful resources directly from the Martian environment. To extract Martian resources, one would need sufficient sources of energy to sustain the needed industry and technology.

An approximation by J. Fogg suggests that an outpost of eight Martians may require 1.2 KWe/individual of energy in the form of electricity, with this trend continuing to increase with an increase in the number of settlers; these requirements may reach a value of 1–100 KWe/individual for a smaller biosphere.[13] In his paper, Fogg estimates the per capita power consumption for a permanent Martian settler

---

[10] NASA JPL, 'Curiosity Rover Finds Biologically Useful Nitrogen on Mars' (*NASA Jet Propulsion Laboratory, 24 march 2015).* www.jpl.nasa.gov/news/news.php?feature=4516. Accessed 25 July 2020.

[11] Rocco L. Mancinelli and Amos Bamin, 'Where is the nitrogen on Mars' (2003) 2(3) International Journal of Astrobiology. https://doi.org/10.1017/S1473550403001599. Accessed 26 July 2020.

[12] Kevin M. Cannon and Daniel T. Britt, 'Feeding One Million People on Mars (2019) 7 The Journal of Space Entrepreneurship and Innovation. https://doi.org/10.1089/space.2019.0018. Accessed 30 July 2020.

[13] Martyn J. Fogg, 'The Utility Of Geothermal Energy On Mars' (1996) 49 JBIS. https://doi.org/10.1089/space.2019.0018. Accessed 4 August 2020.

by assuming consumption of electricity per capita multiplied by a factor reflecting the need to synthesize alternative energy carriers, and any increment further due to needs of artificial life-support.[14]

One of the practical figures for energy per capita came out from Japan-based Ohbayasi corporation, which tried to design and calculate the energy requirements for a permanent base for 150 settlers on Mars. With their approximation, they estimated each person on Mars would roughly require around 20–50 KWe. Electric power requirement on Mars by an individual settler is about 2–5 folds greater than average power consumption per capita on Earth due to the following reasons:

- On Mars, unlike Earth, humans require power to generate O_2 from the Martian environment to ensure living and sustainability on Mars, let alone be the electric power required for the production of H_2O from Martian Poles.
- Unlike on Earth, on Mars, Martian habitats would require a feasible life support system to maintain Pressure, Temperature, Oxygen, and the Humidity level of living quarters. To ensure the proper functioning of these life support systems, a continuous amount of power needs to be supplied to every individual component of the life-supporting system.
- Agriculture could be another aspect to relate power demand on Mars to that of Earth. On Mars, due to insufficient solar Energy, Plant-based cultivation would require supplemental power in terms of LED lighting and extra care for its growth, let alone power requirements for cellular agriculture and insect farming.

A conclusion can be drawn that the average power requirement by an individual Martian settler would be much larger than that of average electric power consumption on Earth. Energy therefore provides a crucial role in limiting the number of people who can live on Mars.

Solar Energy can be a reliable source of energy as here on Earth. However, on Mars, due to its greater distance from the Sun as compared to Earth, the Martian atmosphere hinders all practicality of solar energy. Such obstruction suggests that in the future, a significant fraction of Martian settlers will rely upon fission/fusion based reactors for energy utilization.[15] Nuclear reactors provide the following benefits over other forms of energy sources:

- Nuclear reactors reduce transportation constraints because it requires only half of the weight of material to be transported to that of solar plants to produce similar output.
- Nuclear power plants are much more reliable as compared to other forms of energy sources.
- Unlike other energy sources, nuclear power plants can provide passive heat to warm up settlements in night and seasonal cycles.

---

[14] Ibid.

[15] Mason Black, 'Powering a Colony on Mars' (*Stanford.edu*, 5 December 2017). http://large.stanford.edu/courses/2017/ph240/black1/. Accessed 5 August 2020.

So, the number of people that can live on Mars (N_e) in terms of energy utilization can be calculated by the total number of reactors (N_r) times the average energy supplied by individual nuclear reactors (E) and divided by the average energy required by individuals of Mars (E_a) i.e.

$$N_e = N_r * \left(\frac{E}{E_a}\right)\text{people}$$

Obliviously, there is some uncertainty in calculating and establishing a round figure for the total number of nuclear reactors on Mars. The establishment of such reactors will also depend on geopolitical factors and policies enacted by Martian decision makers. Thus, the number of people that Mars can accommodate in terms of energy factors depends upon the type of power source used to meet a settlement's energy needs, and on the geopolitical factors of Mars.

8.4 Space and Martian Area

Mars has an area of 144.8 trillion m^2. If an attempt is made to fit humans in such a vast area, around 1.5 quadrillion people fit on the Martian surface. For this calculation, consider the area of Mars to be denoted by 'A_m', the land area of Earth be 'A_e', the area required by a single person be 'A_S', and the total number of people that could fit on Mars as 'N', so,

$$10\,\text{humans can fit inside in } 1\,\text{m}^2 \text{ of area}$$

$$A_S = 0.1\,\text{m}^2$$

Also,

$$A_m = 1.44 * 10^{12}\,\text{m}^2$$

$$\text{N} = (A_m/A_S)\text{people} = 1.5\,\text{quadrillion people}$$

However, this number cannot meaningfully apply toward any sustainable living goals on Mars. So, from the above calculation, it is pretty clear that the area alone is not an issue for a Martian settlement.

The first humans may land on Mars by the year 2035 based on some estimates by NASA.[16] It may not be long after that the first human settlement is initiated on

[16]Sean Keach, 'RED PLAN-IT Humans could land on Mars by 2035 thanks to manned Moon mission in five years, Nasa chief claims' (*The Sun,* 22 October 2019). www.thesun.co.uk/tech/10190885/nasa-mars-landing-deadline-date-manned-mission-moon/. Accessed 5 August 2020.

Mars. Probably, the first level of the human settlement would not take place over the Martian surface, as pointed out by Zubrin, due to the radiation exposure and the harsh environment of Mars. Underground Lava tubes may be the best and the feasible option to protect settlers from any temperature variation, micrometeorite bombing, or dust storm in the Martian environment.[17]

However, such an underground settlement would not only trade-off the area per Martian settler, but also would offer Martians a lower opportunity for exploration on the mainland surface. Calculating the population of the first level of a Martian settlement is quite uncertain as of today. Such settlements would not only depend on the discovery of several ideal lava tubes, but also on the Martian regulations and policies which may restrict the total number of people that can settle on Mars.

For successive generations of Martian settlement, the settlement may be more likely to occur on the surface. In such a time frame, the Martian surface may eventually become covered with resources such as living quarters, power sources, ISRU equipment, communication devices, propulsion machines, and scientific research components. With such infrastructure, it has been predicted that the settlers on an average would require more than 200 m^2 of area inside living units to utilize and sustain every Martian resource.[18]

In the model developed by Cannon and Britt, they have predicted that alone the food resources for locally feeding 1 million covers around $4.6 * 10^7 m^2$ of the Martian surface, which does not include the area covered by other ISRU components.[19] With ongoing engineering advancements, it is entirely plausible that future Martian settlements may use a blended method to utilize two or more technological modules under the same region of the Martian surface.

On Mars, it may be possible that settlements will model their ideal population density based on the most sustainable country on Earth to sustain the allocation of Martian resources efficiently. For example, a Martian settlement could consider the population density of Norway, which of as 2020, is considered the most sustainable country on Earth with an approximated Population Density (P.D) of 15 people/km^2, followed by Sweden with 25 people/km^2. So here, estimating a rough figure of the total Martian population based on the population density of Norway and the average population density of Earth's gross population.

(a) For the first case, assuming a Martian population density matches that of Norway:

$$\text{In Norway, 15 people are sustained in 1 km}^2 \text{of area}$$

[17]Sarah Sloat, 'Mars colonists could leave in lava tubes beneath the surface' (*Inverse*, 26 September 2017). www.inverse.com/article/36777-mars-moon-human-colony-lava-tubes. Accessed 5 August 2020.

[18]Mars One, 'Mars One Astronauts' (*Mars One*). www.mars-one.com/mission/mars-one-astronauts. Accessed 8 August 2020.

[19]Supra n 12.

$$A_S = 1/15\,\text{km}^2 = 0.066\,\text{km}^2$$

Also,

$$A_m = 144,371,391\,\text{km}^2$$

So,

$$\text{N} = (144,377,391/0.066)\text{people} = 2.1 \text{ billion people.}$$

(b) For another case, assuming a Martian population density that matches with an average population density of the Earth:

$$\begin{aligned}&\text{Average population density on land area}\\&= (A_e/\text{Total population of Earth})\text{person/km}^2\\&= 52\,\text{person/km}^2\end{aligned}$$

So,

$$A_S = 1/52\,\text{km}^2 = 0.0192\,\text{km}^2$$

$$\text{N} = (144,377,391/0.0192)\text{people} = 7.51 \text{ billion people.}$$

Examining the calculations with the above data, it is seen that the Martian land area is approximately equal to the Earth's land area. Statistics for the Martian population varies according to the intended population density. Also, here, it is fair to assume that on Mars, different states/settlements and private corporations on Mars may solely adopt their population density strategies according to different goals. For example, a U.S.-based organization may adopt sustainable practices that include $\geq \vee \leq 92.9$ settlers per square mile of the Martian land area (current population density of U.S). Similarly, Chinese based settlements may derive sustainable practices which include $\geq \vee \leq 153$ settlers/km^2 (current population density of China) (Fig. 8.2).

With such practices, one can extrapolate the solution that Mars as a whole could conceivably sustain a maximum population close to that of Earth today, approximately 7 billion people.

8.5 Governance Structure

As Martian settlements start to reach self-sustainability, they may develop stable governments to develop and implement necessary laws and regulations. Whatever form of government gets established, such a government would focus on the needs

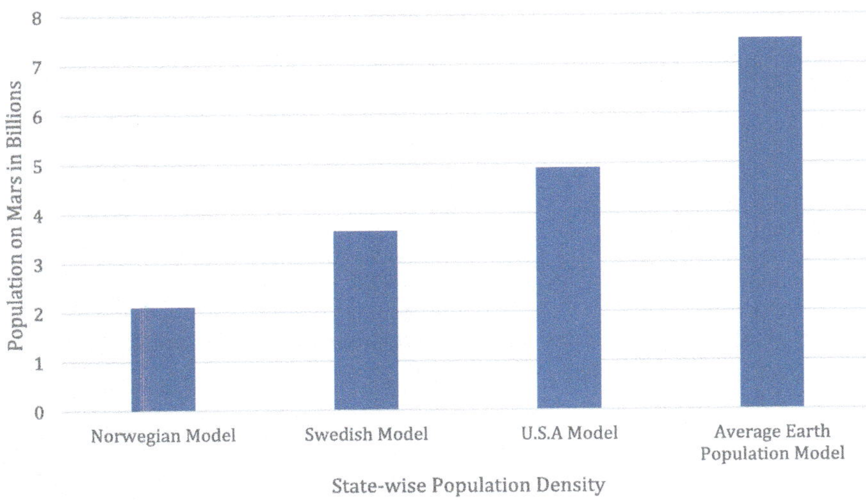

Fig. 8.2 Approximated population density-wise population model for Mars

of Martian settlements than interest in any state on Earth. The type of government thus established on Mars and their related policies could be another limiting factor for determining the Martian population.

For a governance structure, there are several ways by which they could place an upper limit on the number of people that Mars could support. Such an approach may include.

8.5.1 Policies that Restrict the Total Number of People Who Can Live Inside a Settlement

For the first point, it is assumed that a Martian government would take an initiative to prevent the outpacing of Martian resources by limiting the population, the maximum number of people to live inside a habitat. This may be one of the most significant responsibilities of a Martian government, in order to sustain the needs of Martian settlers on Martian land for an extended period. Different corporations or states running different settlements, based on their demographics and policies, might adopt different figures for the maximum number of people that are allowable inside a habitat.

8.5.2 Limiting the Number of People Who Could Be Immigrated from Earth to Mars or Born on Mars, to Prevent Outpacing of the Martian Resources

For the second point, the actions of a Martian government may also restrict the growth of the Martian population by placing a limit on immigration and fertility rate. Any restrictions or limitations on immigration of people from Earth to Mars could be a consequence of tensions between Earth and the emerging Martian governance system or a practical limitation to enable a sustainable use of resources to feed the limited number of Martian settlers.

A Martian governance system may also control the growing population by standardizing and imposing a limited fertility rate to the population of Martian settlers, as a part of their agenda to maintain sustainable conditions for everyone. Since a Martian settlement requires much more focused governance to ensure sustainability and economic goals, a higher fertility rate might threaten the viability of the settlement itself if the population growth outpaces the availability of resources. Different settlements, based on their demographics and sustainability criteria, might also develop their methods for adopting new values regarding fertility rates to their settlers.

8.5.3 Securing Adequate Gender Distribution

Another approach that a Martian government could adopt to limit a growing population is by securing an adequate gender distribution as part of their policy. This could either be an equitable gender distribution, i.e., 50–50 Male to Female ratio and inequitable gender distribution, i.e., 60–40 Male, Female ratio, or vice versa. Ensuring such gender distribution would not only help government structures to project the future needs of the settlement but also would help them to ensure that the growing population on Mars does not outpace the availability of resources.

8.5.4 Prevent the Outpacing of One Age Group Over Another

For the last point, a Martian governance system may restrict the growing population by maintaining a required age distribution as a part of their policy. The age group under 0–14 years is generally considered as the children's age group; people under the age of 15–24 are considered as the youth population; the working-age group is considered under age of 25–64 years, while people of age greater than 64 years are counted as the old age group people. It would likely be people within 25–64 years who would contribute a major role in the Martian development and economy. The Martian government would be sensible about the population age distribution, as too low of a fertility rate and lower immigration of young people from Earth to Mars would result in a shift towards the older age group and hence, a shrinking of the Martian workforce. At the same time, some old age group people would also be required in the Martian population to pass down their culture, tradition, and experience to their descendants. It is also to be noted that Martian governance, perhaps would not be able to sustain an unexpected burst in the population of children and youth, as such age groups would require higher resources and attention than any other age groups. Given the practicality of managing age groups, a Martian government would need to ensure that a particular age group wouldn't outpace another age group in order to avoid situations of resource or labour shortages.

Beside governance structure, there might also be some uncertainties about the Martian settlers' actions, which may force a Martian government to change, develop, and amend their policies. These actions could include:

- Riots or massive public rallies
- Attempts to take down the government
- Denying the act of government policies
- Hunger strikes by Martian public

These examples demonstrate how governance structure and its policies must maintain a careful balance between policies intended to sustain the viability of a settlement and the individual needs of everyone in the Martian population.

8.6 Engineering Structure of Living Quarters

Another factor that needs to be overviewed for assessing the Martian population is the engineering structures of living quarters. A higher number of living quarters on or under the Martian surface would accommodate a larger population. These living quarters on the Martian surface would likely to cover the minimum area required in order to provide a larger volume to lodge a significant number of people. For this same reason, hemispherical—dome (Geodesic dome) like geometry, or an icosahedron shape is likely to be adopted as a living quarter on Mars. They provide the lowest area to volume ratio.[20]

Due to the various engineering advantages of a geodesic-dome shaped structure over icosahedron geometry, it is more likely that the successive Martian generations will consider a geodesic shape to construct their living quarters. A geodesic dome with an adequate thickness will be required to install on Mars, to protect settlers from harmful radiation, and to provide living quarters an extra structural advantage against Martian severe environments.

[20]MrReid.org, 'Spherical ice cubes and surface area to volume ratio' (*MrReid.org*). www.wordpress.mrreid.org/2011/10/20/spherical-ice-cubes-and-surface-area-to-volume-ratio/#:~:text=Of%20all%20the%20Platonic%20solids,well%20suited%20for%20cooling%20drinks. Accessed 10 August 2020.

Martian regolith is abundant with aluminium (Al), titanium (Ti), nickel (Ni), iron (Fe), manganese (Mn). The Viking lander also discovered an abundance of $Si\,O_2$—a vital constituent of glass accounted for 40% of Martian regolith by weight. So, the materials required for constructing such structures are available on Mars. The only thing which may restrict the construction of engineering domes is the technological advancement for ISRU and any guidelines of established governance systems on Mars for the use of resources.

It is more likely that political factors could restrain the construction of such engineering structures even if technology allows to construct habitats. Such political elements may include:

- A prolonged economic recession may enable the government to take some mandatory steps to sustain a limited number of Martian settlers by constructing a limited number of habitats.
- Any settlement efforts by private corporations might be limited by increased tax, interest, and lower subsidies offered by the Martian government.
- Any conservation approach by the Martian government to secure the natural heritage and aesthetics of Mars might also limit the construction of the number of domes all over the Martian surface.

So, besides physical factor like ISRU, political factors could play a significant role in the establishment of such an engineering structure on Mars. These political factors may allow only a maximum or a minimum number of people to stay inside such living quarters. Also, the fundamental sustainability requirements per capita set by different Martian settlements could be another deciding factor to allow a limited number of people to live inside such settlements.

In this way, engineering structures are another essential factor in determining the population to be harbored on the Martian surface.

8.7 Conclusion

There are several uncertainties while estimating the number of people that Mars could accommodate. However, this article described some factors to give insight into the ways they could affect the Martian population. The problem of sustainability on Mars can conceivably be compensated by future technological advancements. From the hierarchical model, it is shown that ISRU advancement could be the most critical factor that affects the Martian population, as, without it, it would be practically impossible for a long-term settlement on Mars. Considering the requirements of Martians for sustainability in terms of energy, land area, political state, and technological advancements, it is illustrated that Mars could hold a population of 2 billion people based on the present-day Norwegian population density. Actual limits on the population may vary, depending upon the capabilities of resource extraction and any guidelines issued by decision makers.

A brief discussion from the article has provided an overview of different kinds of political factors that may place an upper limit to the Martian population. This may suggest skepticism with applying a Norwegian population density to the Martian environment. Indeed, there are many factors that could restrict the Martian population. However, here, a preliminary discussion has been made on some fundamental factors that will affect a future Martian population.

Satyam Tiwari is an undergraduate Aeronautical/Aerospace engineering student from Kurukshetra University, India. During his academic session, he has done his training work with Air India and research work on miniature satellites, model rockets as part of his internship program. Currently, Satyam is pursuing his internship from the Blue Marble Space Institute of Science (BMSIS) on the Mars Settlement and Sovereignty Program.

Chapter 9
Legal Implications for Gender Mixed Human Settlements on Mars—Preliminary Thoughts on Human Reproduction and Childbirth in Space

Claudiu Mihai Tăiatu

Abstract The future human settlements on Mars should not take place in a lawless space and valid international law should frame the activities of human beings on Mars. In this article the analysis of the legal implications of human settlements to Mars refers to the basic rights of human reproduction and childbirth. The perspective of space travel to Mars and the implications of "colonizing" the Red Planet raises legal questions on the possibility of living and even building a family as part of the Mars human settlement. Such perspectives represent a subject that should be brought into consideration because of the implications that results when analyzing the ethical considerations from giving birth in space. Apart from raising medical questions, childbirth on another planet raises also questions of citizenship, governance, property rights and future relationship with legal instruments on Earth. Considering this article as a preliminary research, the analysis will focus on raising some of the most important legal questions on the extension of human rights to the persons on Mars and focusing on childbirth in space.

9.1 Introduction

Addressing women's rights is subject to global commitments. A wide series of international legal instruments have positively impacted political commitments to women's rights and equality over the time. According to the United Nations Human Rights Office of the High Commissioner (OHCHR), "attaining equality between women and men and eliminating all forms of discrimination against women are fundamental human rights and United Nations values".[1] The basic legal instruments

[1]United Nations OHCHR, Women's Rights are Human Rights, 2014. https://www.ohchr.org/Documents/Publications/HR-PUB-14-2.pdf (all websites cited in this article were last accessed and verified on 30 September 2020).

C. M. Tăiatu (✉)
European Space Policy Institute (ESPI), Vienna, Austria
e-mail: c.taiatu@yahoo.com

A. Froehlich (ed.), *Assessing a Mars Agreement Including Human Settlements*, Studies in Space Policy 30, https://doi.org/10.1007/978-3-030-65013-1_9

for women's human rights and equality are twofold, namely the Charter of the United Nations and the Universal Declaration of Human Rights. The Charter of the United Nations was adopted in 1945 and among its goals contains stipulations about the equal rights of men and women, such as the provisions of Article 1 (purpose of the United Nations), Article 13 (mandate of the General Assembly) and Article 55 (promotion of universal human rights). The Universal Declaration of Human Rights was adopted in 1948 and proclaims the equal entitlements of women and men to the rights contained in it.

The lack of scientific information about the life-threatening complications in connection with human pregnancy in space environment makes this study more difficult. However, looking ahead on the possibility of developing human settlements on Mars, such questions should be taken into consideration not only by the space medicine but also by space law and policy. Once a human settlement on Mars will become a reality, should the astronaut women have the right to decide freely and assume the risks of pregnancy and childbirth? or should the activity of reproduction be forbidden? or should studies be available before and taken into consideration? Are we in the stage of space exploration when this problem should be legally considered? What I believe is that discussing this subject will reveal important ethical and legal concerns which have to be discussed bravely prior to a human settlement in space and in particular taken into consideration for planning a space exploration on Mars. With the advent of space tourism missions, it may represent a window of opportunity to discuss the risks of giving birth in space and what legal solutions are available to prevent maternal death and morbidity.

9.2 Selected Space "Firsts" Achieved by Women Astronauts

The first woman to travel into space was the soviet cosmonaut Valentina Tereshkova in 1963.[2]

The first time a space station hosted a mixed-gender crew was in August 1982. On 19 August 1982 cosmonaut Svetlana Y. Savitskaya launched aboard Soyuz T-7 to spend a week aboard the Salyut-7 space station.[3]

The National Aeronautics and Space Administration (NASA) astronaut Sally Ride became the first U.S. woman in space on board the space shuttle Challenger on 18 June 1983 during the STS-7 mission.[4]

---

[2]Moskowitz, C., Women in space: A gallery of firsts, 2019. https://www.space.com/16143-women-space-firsts-gallery.html.

[3]Uri, J., NASA—Space Station 20th—Women and the Space Station. https://www.nasa.gov/feature/space-station-20th-women-and-the-space-station.

[4]Ibid. NASA. https://www.nasa.gov/feature/space-station-20th-women-and-the-space-station.

On 25 July 1984, cosmonaut Svetlana Y. Savitskaya became the first woman to participate in a spacewalk or Extra-Vehicular Activity (EVA) during her second flight to Salyut 7.[5]

The first British woman in space was Helen Sharman, who visited the Mir Space Station in May 1991.

The first African American woman in space was NASA astronaut Mae Jemison in September 1992.

The first Canadian woman astronaut to travel in space was Roberta Bondar in 1992.

The first Japanese woman in space was Chiaki Mukai in July 1994.

The first French woman in space was Claudie Haigneré who flew to the Russian Mir Space Station in 1996.

NASA astronaut Eileen M. Collins was the first woman to command a space shuttle mission in July 1999. Astronaut Eileen M. Collins has previously in 1995 became the first female pilot of a Space Shuttle Mission.

The first woman on the International Space Station was NASA astronaut Susan Helms in March 2001.

The first female space tourist was the Iranian American entrepreneur Anousheh Ansari in 2006.

NASA astronaut Peggy A. Whitson became the first woman to command the International Space Station during Expedition 16 in 2007.[6]

The first South Korean astronaut to go to the International Space Station was Yi So-yeon in April 2008.

Four was the largest number of women serving together on the International Space Station in April 2010, bringing together NASA and JAXA astronauts. From 16 to 23 May 2010 it was the first time when four women were aboard the ISS at one time: Metcalf-Lindenburger, Yamazaki, Wilson and Caldwell Dyson.[7]

The first Chinese woman taikonaut to travel in space was Liu Wang on 19 June 2012 aboard the Shebzhou-9 spacecraft and docking with the Tiangong-1 space station.

The first female ESA astronaut participating on a long-duration mission was Samantha Cristoforetti from Italy.[8] The expedition 42 took place between November 2014 and March 2015.

NASA astronaut Christina Koch spent 328 consecutive days in space from March 2019 to February 2020 and broke the record for the longest single spaceflight by a woman. The most recent record involving women astronauts was in January 2020 when NASA astronauts Christina Koch and Jessica Meir undertook the first all-female EVA/spacewalk.[9]

---

[5] Ibid. NASA. https://www.nasa.gov/feature/space-station-20th-women-and-the-space-station.

[6] Ibid. NASA. https://www.nasa.gov/feature/space-station-20th-women-and-the-space-station.

[7] Ibid. NASA. https://www.nasa.gov/feature/space-station-20th-women-and-the-space-station.

[8] Ibid. NASA. https://www.nasa.gov/feature/space-station-20th-women-and-the-space-station.

[9] Rincon, P., Christina Koch: NASA astronaut sets new female space record. https://www.bbc.com/news/science-environment-51387464.

9.3 Women Leadership in Space Activities

It can be reasonably mentioned that equality between women and men in space activities has been acknowledged since the first woman traveled into space in 1963. The women leadership further developed until the current activities of the International Space Station (ISS). Women are part of future plans for space exploration and could be regarded as indispensable for human settlements on other planets, including Mars. The above-mentioned milestones reveal almost 60 years of space activities when humankind developed its understanding about life support and the consequences of microgravity to woman astronauts. Thus, women leadership in space activities is incontestable and should become more attested for future space travel due to the need for scientific research about the possibility of giving birth in space.

Since 2nd of November 2000 humans have been orbiting our planet continuously within the ISS. In this regard, NASA disclosed that as of March 2020, 65 women have flown in space and of these, 38 women have visited the ISS as long-duration crewmembers, as visitors on Space Shuttle assembly flights or as a Space Flight Participants on short duration Soyuz missions.[10] Most women in space have been U.S. citizens with missions on the Space Shuttle and on the ISS. Also, it could be highlighted that the ISS welcomed the most astronauts by country.[11] NASA revealed that a number of 240 individuals from 19 countries visited the ISS as of 5 June 2020, with the U.S. having the highest number of astronauts (151 people) and the United Arab Emirates (UAE) being the most recent country to send an astronaut to the ISS and preparing to select its first female astronaut.[12] NASA's human lunar exploration plans under the Artemis program call for sending the first woman and the next man to the surface of the Moon by 2024 and establishing sustainable exploration by the end of the decade.[13] The first woman to go the Moon is planned for 2024, as part the U.S. Artemis program.

9.4 The 1969 NASA Study Referring to "Manned" Landing to Mars

The successful landing on the Moon of the Apollo 11 mission and the enthusiasm for space travel contributed to the establishment of a new Space Task Group formed to recommend a post-Apollo manned space program. The Space Task Group was a NASA working group created in 1958 and tasked with managing the U.S. human spaceflight programmes. The Space Task Group was later transformed into the

[10]Ibid. NASA. https://www.nasa.gov/feature/space-station-20th-women-and-the-space-station.

[11]Garcia, M., NASA—Visitors to the Station by Country, 5 June 2020. https://www.nasa.gov/feature/visitors-to-the-station-by-country/.

[12]Nasir, S., 1,400 Emirati women apply to be UAE's next astronaut. https://www.thenational.ae/uae/science/1-400-emirati-women-apply-to-be-uae-s-next-astronaut-1.1017405.

[13]https://www.nasa.gov/topics/moon-to-mars/overview.

Manned Spacecraft Center and now the Lyndon B. Johnson Space Center located in Houston, Texas. On 4 August 1969, Wernher von Braun briefed the Space Task Group with the plan for a manned expedition to Mars.[14] When reading the von Braun's study of "Manned Mars Landing Presentation to the Space Task Group" highlights that:

> With the recent accomplishment of the manned lunar landing, the next frontier is manned exploration of the planets. (…) The information presented here describes a method of landing men on the planet Mars in 1982. (…) A 1982 manned Mars landing is a logical focus for the programs of the next decade.[15]

This plan refers to six people per ship, and the deployment of two ships for a 270-day journey to Mars and an 80-day period on the surface of Mars. This mission was designed to carry three men on each ship to the surface of Mars and return the crew, scientific data and samples to the spaceship. The study presented by von Braun to NASA Space Task Group refers to "manned" Mars landing mission. Without a doubt, reference is intentionally given to man/men without including reference to woman/women. For example: "the Earth orbital program will qualify both man and systems for long duration while the lunar program will provide techniques and experience for surface operations".[16] Also, the last figure containing the conclusion of von Braun's presentation to NASA's Space Task Force include a plan for 1970–1990 referring to the necessary number of men envisioned for space missions to the Geosynchronous station, in Low Earth Orbit, in Lunar Orbit, on Lunar Surface, on the temporary base and semi-permanent base on Mars.[17] The NASA Space Task Group final report on 15 September 1969 approved von Braun's plan and recommended all the infrastructure to Mars. In the end, this plan was not approved by the White House, the U.S. President Nixon budget office has only approved the space shuttle design, without approving plans for Mars.[18]

---

[14]von Braun Mars Expedition, 1969. http://www.astronautix.com/v/vonbraunmarpedition-1969.html.

[15]NASA, von Braun, Wernher "Manned Mars Landing, Presentation to the Space Task Group", 4 August 1969, pp. 2. https://www.nasa.gov/sites/default/files/atoms/files/19690804_manned_mars_landing_presentation_to_the_space_task_group_by_dr._wernher_von_braun.pdf.

[16]NASA, von Braun, Wernher "Manned Mars Landing, Presentation to the Space Task Group", 4 August 1969, pp. 26. https://www.nasa.gov/sites/default/files/atoms/files/19690804_manned_mars_landing_presentation_to_the_space_task_group_by_dr._wernher_von_braun.pdf.

[17]NASA, von Braun, Wernher "Manned Mars Landing, Presentation to the Space Task Group", 4 August 1969, pp. 51. https://www.nasa.gov/sites/default/files/atoms/files/19690804_manned_mars_landing_presentation_to_the_space_task_group_by_dr._wernher_von_braun.pdf.

[18]Ibid., von Braun Mars Expedition, 1969. http://www.astronautix.com/v/vonbraunmarpedition-1969.html.

9.5 The U.S. Policies Referring to Human Space Travel and Including Mars Missions

The U.S. President Barack Obama issued on 28 June 2010 a National Space Policy directive.[19] The 2010 "National Space Policy of the United States of America" contains relevant provisions for the exploration of Moon and Mars. In particular, provides that the NASA Administrator shall begin crewed missions beyond the Moon, including sending humans to an asteroid while by the mid-2030 s to send human to orbit Mars and return them safely to Earth.[20]

The U.S. President Donald Trump issued on 11 December 2017 the Presidential Memorandum on "Reinvigorating America's Human Space Exploration Program" known as the Space Policy Directive-1 (SPD-1).[21] SPD-1 amended elements of the 2010 National Space Policy, requesting a sustainable program of space exploration, the return of humans to the Moon, followed by human missions to Mars.[22]

As a result of the provisions under Space Policy Directive-1, NASA announced on 13 December 2018 that it seeks U.S. partners to develop reusable systems to land astronauts on Moon, with the goal of sending crew to the surface in 2028.[23] According to such plans, NASA envisioned to send crewed missions to the Moon not earlier than 2028. However, in March 2019 during the meeting at the National Space Council, the U.S. Vice President Mike Pence presented the policy of the U.S. to return American astronauts to the Moon in 2024, four years earlier than originally planned. This 2019 request represented a major acceleration for the NASA programs in comparison with the original NASA plans intended to landing humans on the Moon by 2028.[24] Such a new perspective also accelerates the possibility to send humans to Mars.

---

[19] Office of Space Commerce, National Space Policy, 2010. https://www.space.commerce.gov/policy/national-space-policy/.

[20] National Space Policy of the United States of America. https://obamawhitehouse.archives.gov/sites/default/files/national_space_policy_6-28-10.pdf.

[21] White House, Presidential Memorandum on Reinvigorating America's Human Space Exploration Program, 11 December 2017. https://www.whitehouse.gov/presidential-actions/presidential-memorandum-reinvigorating-americas-human-space-exploration-program/.

[22] White House, Presidential Memorandum on Reinvigorating America's Human Space Exploration Program, 11 December 2017. https://www.whitehouse.gov/presidential-actions/presidential-memorandum-reinvigorating-americas-human-space-exploration-program/.

[23] NASA, NASA seeks US Partners to Develop Reusable Systems to Land Astronauts on Moon. https://www.nasa.gov/feature/nasa-seeks-us-partners-to-develop-reusable-systems-to-land-astronauts-on-moon/.

[24] Foust, J., Pence calls for human return to the moon by 2024, March 2019. https://spacenews.com/pence-calls-for-human-return-to-the-moon-by-2024/.

9.6 NASA Artemis Plan to Land First Woman and Next Man on Moon in 2024

In September 2020, NASA published an update of the Artemis program. The "Artemis Plan, NASA's Lunar Exploration Program Overview" is structured in three chapters and contains seven appendixes. As a foreword, NASA Administrator Jim Bridenstine highlights the coalition of partnerships which NASA is building with industry, nations and academia. NASA Administrator highlighted the activities under Artemis programme to explore more of the Moon than ever before "uniting people around the unknown, the never seen, and the once impossible".[25] This quote describes the space exploration plans in which the humankind prepares to adventure itself and highlights the continuous learning process for these operations. According to the NASA update on the Artemis plans, the Artemis I and Artemis II missions will prepare the future missions and provide performance data and operational experience while the Artemis III is intended to land the first woman and next man on the Moon surface in 2024.[26]

9.7 The Outer Space Treaty Provisions, International Law and United Nations Instruments

The 1967 Treaty on "Principles Governing the Activities of States in the Exploration and Use of Outer Space, Including the Moon and Other Celestial Bodies" (OST) contains provisions about the applicability of international law, including the Charter of the United Nations in the exploration and use of outer space. The provisions of Article III OST stipulate that activities in outer space must be carried out in accordance with international law, including the UN Charter, in the interest of maintaining international peace and security and promoting international cooperation and understanding.[27] It has been underlined in academic literature that this OST principle could be considered as customary law, meaning that such provisions are binding and relevant for all states with space based activities.[28]

International law is enshrined in conventions, treaties and standards and many of the United Nations (UN) legal instruments govern the relations among nations.[29] In particular, international public law encompasses the UN system and bilateral or multilateral intergovernmental treaties or organizations.[30]

[25]NASA, Artemis Plan. NASA's Lunar Exploration Program Overview, September 2020. https://www.nasa.gov/sites/default/files/atoms/files/artemis_plan-20200921.pdf.

[26]NASA, Artemis Plan, NASA's Lunar Exploration Program Overview, September 2020, pp. 20. https://www.nasa.gov/sites/default/files/atoms/files/artemis_plan-20200921.pdf.

[27]Art. III OST.

[28]Lyall, F., Larsen, P.B., Space Law, 2009, pp. 510.

[29]United Nations, International Law and Justice. https://www.un.org/en/sections/issues-depth/international-law-and-justice/index.html.

[30]White, H.M., International Law and Relations. https://er.jsc.nasa.gov/seh/law.html.

The UN system being part of the OST is relevant to human rights and especially to women reproductive rights. Reproductive rights have been considered as human rights within the UN system because the UN instruments protect elements of reproductive rights, however there are no instrument specifically dedicated to reproductive rights.[31]

The European Court of Human Rights decisions on matters relating to reproductive rights, the finding that the right to private life incorporates the right to respect the decisions both to become and not to become a parent in the Judgment by the Grand Chamber of the European Court of Justice, Evans versus the UK 6339/05, (10 April 2007), para. 71. The European Court of Human Rights acknowledged that the right to decide whether to have children is contained within the interpretation and sphere of the right to private and family life in Article 8 of the ECHR:

> The Court recalls that the notion of "private life" within the meaning of Article 8 of the Convention is a broad concept which encompasses, inter alia, the right to personal autonomy and personal development (see Pretty v. the United Kingdom, cited above, § 61). It concerns subjects such as gender identification, sexual orientation and sexual life (for example, Dudgeon v. the United Kingdom, judgment of 22 October 1981, Series A no. 45, pp. 18–19, § 41; and Laskey, Jaggard and Brown v. the United Kingdom, judgment of 19 February 1997, Reports of Judgments and Decisions 1997-I, p. 131, § 36), a person's physical and psychological integrity (Tysiąc v. Poland judgment, cited above, § 107) as well as decisions both to have and not to have a child or to become genetic parents (Evans v. the United Kingdom [GC], cited above, § 71). Case of A, B, C v. Ireland, Application no. 25579/05.[32]

Also, paragraph 7.3 of the International Conference on Population and Development (ICPD) Programme of Action mentions that reproductive rights are based on the right of couples and individuals to decide free from discrimination, coercion and violence whether to have children, how often and when to do so, having the necessary information and means to make such decisions.[33] The 1994 ICPD was endorsed by the United Nations General Assembly Resolution A/RES/49/128.

9.8 The First Private Mission Intended to Demonstrate the Possibility of Giving Birth in Space Was Suspended in September 2019

Delivering a healthy child in microgravity was never proved possible within the current scientific research and with current advancements in space medicine. In January 2019, the start-up "SpaceLife Origin" expressed its intention to send a pregnant woman to space to give birth. Such mission was explained as an experiment in

[31] OHCHR, Reproductive Rights are Human Rights, pp. 21. https://www.ohchr.org/documents/publications/nhrihandbook.pdf.

[32] Ibid. OHCHR, pp. 107. https://www.ohchr.org/documents/publications/nhrihandbook.pdf.

[33] Ibid. OHCHR, pp. 21. https://www.ohchr.org/documents/publications/nhrihandbook.pdf.

space. Within the scope of the mission it was highlighted that planning human settlements without learning how to reproduce in space is "pointless" and thus, the plan was to have a space mission for a "spacefaring childbirth" in 2024.[34] In a mission update in September 2019, it was mentioned that all official operational activities of SpaceLife Origin are suspended.[35] The reasons for suspending the activity included medical, safety and ethical concerns:

> "Serious ethical, safety and medical concerns related to the specific Missions Lotus & Cradle are preventing me any longer from accepting any associations with and responsibilities for those two specific Missions involving 'embryos, pregnant women and baby's in space'. In short: "Better safe than sorry" so I need to distance myself from these missions".[36]

9.9 International Space Station (ISS) and Medical Research

Space technologies from the ISS and know-how are constantly applied back to Earth in the so-called spin-off rewards. For space agencies the most important is that these rewards are directed to public benefit, spin-off means that space technology can benefit society as a result of the money invested in space agency's work.[37] The European Space Agency (ESA) revealed in 2019 that over 700 new start-ups have been fostered in Europe at the 20 ESA Business Incubation Centres in 16 European countries, all based on space spin-offs. Officers from JAXA, ESA, NASA, CNES, CSA, ASI and ISA presented studies on the space technology transfer themes.[38]

The ISS is considered as a unique laboratory for performing investigations that affect human health both in space and on Earth.[39] As part of space research, the benefits for medicine include advances in telemedicine, disease models, psychological stress response systems, nutrition, cell behavior and environmental health.[40] Studies in space medicine include development of robotic arms, preventing bone loss and kidney stones in space flight, high-quality protein crystal growth, vaccine development, advance ultrasound, early detection of immune changes, cancer treatment, water purification, etc.[41] Thus, it results that medical experiments are necessary to

---

[34] Koren, M, The Atlantic—Imagine Giving Birth in Space, January 2019. https://www.theatlantic.com/science/archive/2019/01/space-childbirth-babies/579064/.

[35] https://spacelifeorigin.com/en.

[36] https://spacelifeorigin.com/en.

[37] Simpson, M., Spin-out and Spin-in in the Newest Space Age, 2010. https://iisc.im/wp-content/uploads/2016/07/Spin-Out-and-Spin-In-Simpson.pdf.

[38] ESA, Seven space agencies presented space spin-off potentials. https://www.esa.int/Applications/Telecommunications_Integrated_Applications/Technology_Transfer/Seven_space_agencies_presented_space_spin-off_potentials.

[39] International Space Station, Benefits for Humanity, pp. ix. https://www.nasa.gov/pdf/626862main_ISS_Benefit_for_Humanity.pdf.

[40] NASA, Overview of ISS Research Benefits to Human Health. https://www.nasa.gov/mission_pages/station/research/benefits/human_health.html.

[41] NASA, Overview of ISS Research Benefits to Human Health. https://www.nasa.gov/mission_pages/station/research/benefits/human_health.html.

understand human body reactions in space and that such experiments will continue to be carried on for the benefit of society and space travel.

9.10 Ethics and Human Rights

According to the OHCHR multiple human rights relate to women's sexual and reproductive health, including the right to life; the right to be free from torture; the right to health; the right to privacy; the right to education; and the prohibition of discrimination.[42] Women's right to health includes their sexual and reproductive health, which implies State obligations to respect, protect and fulfill rights related to women's sexual and reproductive health.[43]

OHCHR issued the report to assist policymakers in improving women's health and rights "by providing guidance on devising, implementing and monitoring policies and programmes to reduce maternal mortality and morbidity, and fostering accountability in accordance with human rights standards".[44] According to this report, international human rights law includes fundamental commitments of States to enable women to survive pregnancy and childbirth. Recognition of the right to health include sexual and reproductive health.[45] Maternal mortality and morbidity was qualified as a product of discrimination against women, and denial of their human rights, including sexual and reproductive health rights.[46]

Human rights require particular attention to vulnerable or marginalized groups. From a selected point of view such as the distance to Earth and special conditions of the environment, pregnant women and children living on Mars could be part of vulnerable groups and thus special protection would be necessary. According to the OHCHR Report, states should protect against interference with sexual and reproductive health rights by third parties by enforcing appropriate laws, policies, regulations and guidelines. Thus, States would be responsible for "exercising due diligence, or acting with a certain standard of care, to ensure that non-governmental actors, including private service providers, insurance and pharmaceutical companies, and manufacturers of health-related goods and equipment, as well as community and family members, comply with certain standards".[47] If they fail to act with due diligence to prevent, investigate and punish violations of human rights, States may be held responsible even for private acts.[48]

---

[42]OHCHR, Sexual and reproductive health and rights. https://www.ohchr.org/EN/Issues/Women/WRGS/Pages/HealthRights.aspx.

[43]OHCHR, Sexual and reproductive health and rights. https://www.ohchr.org/EN/Issues/Women/WRGS/Pages/HealthRights.aspx.

[44]OHCHR. https://www2.ohchr.org/english/issues/women/docs/A.HRC.21.22_en.pdf.

[45]OHCHR, para 12. https://www2.ohchr.org/english/issues/women/docs/A.HRC.21.22_en.pdf.

[46]OHCHR, para. 14. https://www2.ohchr.org/english/issues/women/docs/A.HRC.21.22_en.pdf.

[47]OHCHR, para 22. https://www2.ohchr.org/english/issues/women/docs/A.HRC.21.22_en.pdf.

[48]OHCHR, para 22. https://www2.ohchr.org/english/issues/women/docs/A.HRC.21.22_en.pdf.

The right of women to take decisions that affect their lives, including sexual and reproductive health should be analyzed from both the perspectives, including the decision to have children in special conditions. Many women globally experience poor treatment during childbirth, including abusive, neglectful, or disrespectful care. Every woman has the right to dignified, respectful sexual and reproductive health care, including during childbirth. A human rights-based approach for human settlements on Mars depends on fostering accountability of multiple actors and at various levels, within and beyond the health sector including the States and regulatory authorities.[49]

> "Eventually, people living in space could evolve to be different enough from people on Earth that we would consider them to be different species".[50] Without the necessary technology such as artificial gravity, it may be raised the question of unethical human experimentation that violates the principles of medical ethics. Studies about neglectful, abusive and disrespectful treatment of women during childbirth and mistreatment of women during childbirth across all geographical and income-level settings are carried out by UN entities. When referring to women's rights to dignified and respectful care, the World Health Organization (WHO) underlined the need to ensure universal access to safe, acceptable, good quality sexual and reproductive health to reduce maternal morbidity and mortality.[51]

9.11 Is the Informed Consent Procedure Applicable to Childbirth in Space?

Information about childbirth in space is only available as scientific assumptions because we can rigorously accept that it never happened such as the use of in vitro fertilization for a woman to remain pregnant in space. Assumptions are that such activity would be dangerous to the humankind, both for the women and also for the child, due to microgravity effects on human body and the radiation effects in space.

Delivering a child in microgravity may bring a series of medical and legal challenges. A risk of giving birth in low or microgravity is the ectopic pregnancy, while the higher radiation levels raise the probability of birth defects.[52] Would it be possible to cover these risks under the informed consent agreement? Would any authority license the activity to send a pregnant woman to space to give birth or to become pregnant and give birth in space? Could be regarded as a criminal act?[53] This are

---

[49]OHCHR, para 75. https://www2.ohchr.org/english/issues/women/docs/A.HRC.21.22_en.pdf.

[50]Kim, G., Wilkin, R., If humans gave birth in space, babies would have giant, alien-shaped heads. https://www.businessinsider.com/humans-gave-birth-space-earth-giant-alien-heads-2019-7.

[51]World Health Organization, The prevention and elimination of disrespect and abuse during facility-based childbirth. https://apps.who.int/iris/bitstream/handle/10665/134588/WHO_RHR_14.23_eng.pdf;jsessionid=5AF1D217DDF722FC89CC109D9290D13A?sequence=1.

[52]Kramer, M., Sex in Space Could be Out of this World. Or Not, See also: Laura Woodmansee, "Sex in Space", April 2013. https://www.space.com/20718-sex-childbirth-space-mars-mission.html.

[53]Koren, M., Imagine Giving Birth in Space. https://www.theatlantic.com/science/archive/2019/01/space-childbirth-babies/579064/.

questions that may be raised, and it would become useful to analyze the possibility to use the informed consent for licensing such activities.

9.11.1 Informed Consent for Medical Investigations and Treatment

Both the practice and the legal requirement is to use the informed consent agreement for medical investigations, treatment and procedures. The informed consent means that the medical doctor will inform and explain the patient the diagnosis and the different treatment options with their advantages, disadvantages and achievable outcomes.[54] By receiving an informed consent, the patient will decide what treatments (s)he wants to receive. The informed consent was regarded as a "collaborative decision-making process" which includes both an ethical and a legal obligation from the healthcare providers.[55]

Informed consent in medical practice is closely linked to philosophical notions of respect for persons and for individual autonomy. Elements of informed consent include: (i) voluntarism—the patient should be free to decide; (ii) capacity—the patient should have the ability to make health care decisions; (iii) disclosure—the patient should have all the information to understand the medical procedure, including the nature and purpose of the treatment, its risks, potential benefits, and available alternatives; (iv) understanding—the patient can comprehend the information that was provided by the medical doctor; (v) decision—the patient would have the right to authorize or deny the to the physician to execute the proposed treatment.[56]

9.11.2 Informed Consent for Medical Trials on Human Subjects

The consolidated text of the Directive 2001/20/EC "on the approximation of laws, regulations and administrative provisions of the Member States relating to the implementation of good clinical practice in the conduct of clinical trials on medicinal products for human use" scope is to establish specific provisions regarding the conduct of clinical trials on human subjects involving medical products.[57] The provisions of

---

[54] Selinger, C.P., The right to consent: Is it absolute?, 2009. https://www.bjmp.org/content/right-consent-it-absolute.

[55] Healthline. https://www.healthline.com/health/informed-consent.

[56] Joffe, S., del Carmen, M.G., Informed Consent for Medical Treatment and Research: A Review. https://doi.org/10.1634/theoncologist.10-8-636.

[57] Directive 2001/20/EC, OJ L 121, 1.5.2001, pp. 34–44, consolidated version 2009. https://eur-lex.europa.eu/legal-content/EN/TXT/?uri=CELEX:02001L0020-20090807.

Article 2 defines "informed consent" and establishes an obligation to establish an "Ethics Committee" to give its opinion before a clinical trial commences.

9.11.3 *Informed Consent for Space Tourism*

The informed consent role is to release the operator from liability. The informed consent agreement is used in space activities, in particular to organize spaceflights for space tourists. The application of the informed consent that would presumably release the operator from liability towards the participants in case of accident during the flight was analyzed in relation to space tourism, but this was regarded that by no means is a settled matter.[58]

9.12 Conclusions

Currently, human settlements on Moon and Mars could be regarded as projects of experimental nature. It was not yet tested in outer space and implicitly not yet declared safe and operational any of the technologies that would allow a human being to live on another planet than Earth. Until such technology would be available, the idea of childbirth in space remains a pure theoretical question that may not become licensed at least for ethical concerns.

However, becoming pregnant in space may happen in exceptional circumstances. The consequences may result whether in a criminal offence, or just a civil breach of contract depending on the interpretation of law. What is certain is that the high risk of danger combined with the likelihood of medical complications raise ethical, moral and legal challenges reason for which an informed consent agreement may not to be enough to allow a childbirth to take place in space. To balance the preventable maternal deaths, health complications and infant mortality, the decision of having children on Mars should be taken based on scientific research but until such research may be allowed, the technology of traveling and living on Mars should be demonstrated first.

Claudiu Mihai Tăiatu is a Romanian lawyer, he graduated from the Adv. LL.M. of Air and Space Law of the International Institute of Air and Space Law (IIASL), Leiden University, The Netherlands and from the International Space University (ISU), Space Studies Program (SSP18). He successfully completed several internships at the Regulatory Affairs Department of OneWeb, the International Telecommunication Union (ITU) Radiocommunication Bureau, ESPI

[58]Moro-Aguilar, R., National Regulation of Private Suborbital Flights: A Fresh View, pp. 690. https://ecollections.law.fiu.edu/cgi/viewcontent.cgi?referer=https://www.google.com/&httpsredir=1&article=1265&context=lawreview.

and UNIDROIT. Currently, he is working with the European Space Policy Institute (ESPI) in projects related to space law and policy.

Chapter 10
Who Speaks for Mars? The Responsibility to Protect and the Search for Life

André Siebrits

Abstract Humans have always been fascinated by the Red Planet, and by the prospect of life existing beyond the Earth. While the search for life on Mars has primarily focused on finding evidence for past life, the intriguing possibility remains that life, however simple, may still survive and thrive there, and the recent (albeit disputed) discovery of phosphine gas in the atmosphere of Venus has presented new hopes in this regard. Human exploration of outer space, including Mars, is guided by a regime comprised primarily of five core United Nations treaties making up the bulk of the corpus of international space law, but concerns exist regarding the ability of this law to keep up with the new challenges posed by the increasing number of space agencies, including commercial space actors, and their activities in relation to issues of non-appropriation and non-contamination, among others. The central concern of this chapter is with the legal basis of preserving life, if found, on Mars, especially given the proposals for establishing and increasing a human presence there in the coming decades. While planetary protection and non-contamination provisions have generally been given serious attention, are they enough to protect life on Mars if it exists? Looking towards examples on Earth of legal rights granted to natural features and biomes (such as rivers and forests) may provide a fruitful avenue for humanity's future engagement with Mars.

10.1 Introduction

> What account would we give of our stewardship of the planet Earth? We have heard the rationales offered by the nuclear superpowers. We know who speaks for the nations. But who speaks for the human species? Who speaks for Earth? ... Our loyalties are to the species and the planet. *We* speak for Earth.
>
> Carl Sagan, *Cosmos* (Ballantine Books 2013) 347, 365.

A. Siebrits (✉)
Department of Political Studies, University of Cape Town, Rondebosch, South Africa
e-mail: SBRAND003@myuct.ac.za

A. Froehlich (ed.), *Assessing a Mars Agreement Including Human Settlements*, Studies in Space Policy 30, https://doi.org/10.1007/978-3-030-65013-1_10

On Earth, the Responsibility to Protect (R2P) is a global political commitment stemming from the failure of the international community to quickly halt the atrocities committed in the Balkans and Rwanda during the 1990s.[1] If governments and militaries could hide their atrocities behind an inviolable concept of sovereignty, then how could the international community act to prevent similar crimes in future? The view that sovereignty entails not just a right to non-interference but also a responsibility to the wellbeing of a state's residents became formalised as R2P in 2001 following a report by the International Commission on Intervention and State Sovereignty (ICISS), supported by the Canadian government, which espoused the idea of 'states having positive responsibilities for their population's welfare, and to assist each other', and while each individual state has the primary responsibility with regards to its own population, there is nevertheless a 'residual responsibility' vested in the 'broader community of states, which is "activated when a particular state is clearly either unwilling or unable to fulfil its responsibility to protect or is itself the actual perpetrator of crimes or atrocities"'.[2] In 2005, at a high-level United Nations (UN) summit, all UN member states endorsed the fundamental responsibility to prevent and halt, through the UN system, the 'most serious violations of international human rights and humanitarian law'.[3] The emergence of this new international norm represented a significant modification of the longstanding principle of absolute state sovereignty. But what does R2P have to do with Mars and the search for life there?

Recently, the idea has been put forth that R2P should be expanded and applied to the natural world, in terms of the responsibility to protect nature.[4] After all, all humans on the planet share the same biosphere, breathe the same air, and so forth, so should there not be a responsibility to act if countries ignore or expressly support '[u]nconscionable destruction' of the environment?[5] While this idea certainly is worthy of consideration, many states have embarked upon a new path in their relationship with the environment, as seen by the granting of legal rights to natural objects, features, and biomes. One such example is the establishment of the Colombian Amazon as an 'entity subject of rights' by the highest court of that country.[6] More examples were discussed in a previous publication related to the protection

[1] United Nations Office on Genocide Prevention and the Responsibility to Protect, 'Responsibility to Protect' (2020). https://www.un.org/en/genocideprevention/about-responsibility-to-protect.shtml. Accessed 15 October 2020.

[2] Ibid.

[3] Ibid. See also A/RES/60/1: https://www.un.org/en/development/desa/population/migration/generalassembly/docs/globalcompact/A_RES_60_1.pdf.

[4] Juuso Järviniemi, 'The case for a "responsibility to protect nature"' (*The New Federalist*, 9 September 2019). https://www.thenewfederalist.eu/the-case-for-a-responsibility-to-protect-nature?lang=fr. Accessed 15 October 2020.

[5] Ibid.

[6] Anastasia Moloney, 'The Colombian Amazon now has the same legal rights as you' (*World Economic Forum*, 10 April 2018). https://www.weforum.org/agenda/2018/04/colombias-top-court-orders-government-to-protect-amazon-forest-in-landmark-case/. Accessed 15 October 2020.

of cultural heritage sites on the Moon.[7] This is a key concept that will be explored later in this chapter as well. However, to return to the relevance of these ideas to a Mars agreement and future human settlement on the red planet, the recent (but disputed) detection of a potential biosignature—phosphine gas—in the atmosphere of Venus has ignited excited debate across the globe regarding the possibility that life might exist in a relatively mild zone of the Venusian atmosphere.[8] Only life is currently known to produce phosphine, and in the case of Venus, atmospheric anaerobic microbial organisms may be the cause. While this is not proven as of the writing of this chapter the possibility of life, however simple, surviving and persisting in the harsh environment of Venus increases the prospects of finding similarly resilient organisms on Mars. This raises the central question of this chapter—from a legal standpoint, what responsibility do we humans have to protect such life should we encounter it during our exploration and settlement of Mars and the inevitable subsequent exploitation of its natural resources? What happens when our human dream to become a multiplanetary species places our needs in competition with those of pre-existing life, however simple it may be—the same life we have been seeking out for so much of our history? Whose interests will prevail and how can international law adapt to reconcile the interests of humans and life native to other worlds?

10.2 The Current State of the Law

> [One philosopher] asserted that he knew the whole secret … [H]e surveyed the two celestial strangers from top to toe, and maintained to their faces that their persons, their worlds, their suns, and their stars, were created solely for the use of [M]an. At this assertion our two travelers let themselves fall against each other, seized with a fit of … inextinguishable laughter.
>
> Voltaire, *Le Micromégas* (1752) quoted in Carl Sagan, *Pale Blue Dot: A Vision of the Human Future in Space* (Ballantine Books 1994) 20.

Human exploration and expansion into outer space is guided by a core set of international treaties and UN General Assembly resolutions, and while there is no central authority to enforce these rules given the anarchical nature of the interstate system, there has nevertheless emerged a space regime which acts as the guiding force directing humanity's activities in space. Of the five treaties, the Outer Space Treaty (OST), which entered into force in 1967, enjoys the greatest support and, as

[7] André Siebrits, 'The Moon that Owns Itself: Exploring New Legal Avenues to Protect Cultural and Natural Heritage in Space' in Annette Froehlich (ed), *Protection of Cultural Heritage Sites on the Moon* (Springer 2020).

[8] Shannon Stirone, Kenneth Chang and Dennis Overbye, 'Life on Venus? Astronomers See a Signal in Its Clouds' (*New York Times*, 14 September 2020). https://www.nytimes.com/2020/09/14/science/venus-life-clouds.html. Accessed 15 October 2020.

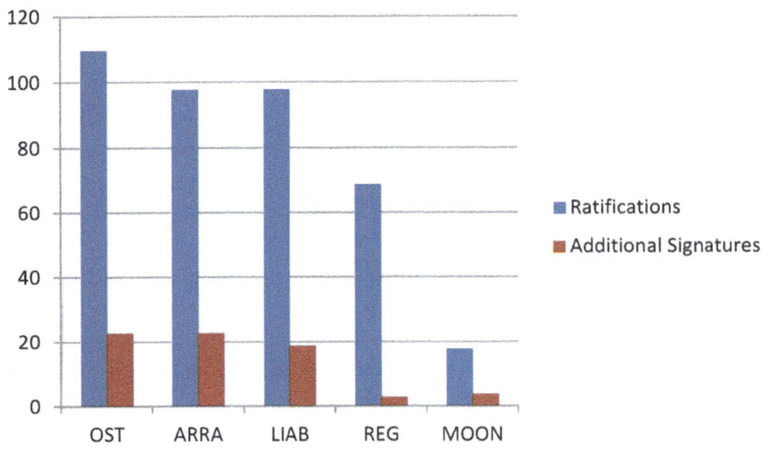

Fig. 10.1 Ratification/accession of Outer Space Treaty (OST), Rescue Agreement (ARRA), Liability Convention (LIAB), Registration Convention (REG), and Moon Treaty (MOON), plus additional signatures as of 1 January 2020 (United Nations Office for Outer Space Affairs, Status of International Agreements Relating to Activities in Outer Space as at 1 January 2020 (2020). https://www.unoosa.org/documents/pdf/spacelaw/treatystatus/TreatiesStatus-2020E.pdf. Accessed 15 October 2020)

of 1 January 2020, 110 states have ratified or acceded to the OST.[9] Beyond these, there are a further 23 states which have signed the treaty. Figure 10.1 depicts the relative levels of acceptance of the five main outer space treaties, demonstrating the critical role of the OST as the 'Magna Carta' of outer space law.

Except for the Moon Treaty, the other treaties all enjoy the support of the major space faring states, while only France has signed but not ratified the Moon Treaty. Accordingly, the first step in the analysis of space law must be the OST, which is also the most comprehensive space treaty. It is well-known that Articles I and II of the OST establish that outer space, including celestial bodies such as the Moon and planets, are the 'province of all mankind' and not subject to national appropriation.[10] In these articles, the rights of humankind to explore and 'use' outer space and the celestial bodies are established and further articles such as Art. III and IV emphasise that all human activities should be peaceful in nature and be undertaken in a spirit of cooperation.

Only in Art. IX is mention made of humankind's responsibilities towards this province it has claimed: 'States Parties to the Treaty shall pursue studies of outer space, including the Moon and other celestial bodies, and conduct exploration of

[9]United Nations Office for Outer Space Affairs, 'Status of International Agreements Relating to Activities in Outer Space as at 1 January 2020' (2020). https://www.unoosa.org/documents/pdf/spacelaw/treatystatus/TreatiesStatus-2020E.pdf. Accessed 15 October 2020).

[10]United Nations Office for Outer Space Affairs, *International Space Law: United Nations Instruments* (United Nations 2017) 4.

them so as to avoid their harmful contamination and also adverse changes in the environment of the Earth resulting from the introduction of extraterrestrial matter and, where necessary, shall adopt appropriate measures for this purpose'.[11] Here lies the crux of this chapter—given this stipulation, what can be considered as 'appropriate measures' to protect (in this case) the Martian environment from the harmful contamination and adverse environmental changes which grow in likelihood in proportion to the increase in human activities on the planet? After all, it was concluded after consideration that

> there is no definition as to what constitutes harmful contamination, nor does the treaty specify under what circumstances it would be necessary to "adopt appropriate measures" or which measures would in fact be "appropriate", while unofficial legal reviews argued that "harmful" should be interpreted as "harmful to the interests of other states," and since "states have an interest in protecting their ongoing space programs," Article IX must mean that "any contamination which would result in harm to a state's experiments or programs is to be avoided".[12]

Additionally, neither the Rescue Agreement, the Liability Convention (which defines damage only in human-centric terms in Art. I as 'loss of life, personal injury or other impairment of health; or loss of or damage to property of States or of persons, natural or juridical, or property of international intergovernmental organizations'[13]), nor the Registration Convention add any further stipulations with regards to protecting planetary environments in outer space. The Moon Treaty, which applies to Mars (and all celestial bodies) as well, does however expand on the OST by stipulating in Art. V-3 that 'States Parties shall promptly inform the Secretary-General, as well as the public and the international scientific community, of any phenomena they discover in outer space, including the Moon, which could endanger human life or health, *as well as of any indication of organic life*'.[14] Here we see the first recognition given to life which may be found in outer space, as well as a marked shift in tone from Art. V of the OST: 'States Parties to the Treaty shall immediately inform the other States Parties to the Treaty or the Secretary-General of the United Nations of any phenomena they discover in outer space, including the Moon and other celestial bodies, which could constitute a danger to the life or health of astronauts'.[15] Here, a public announcement is contingent on the estimation of the threat posed to astronauts only, and not to the Earth as a whole, and certainly not by the astronauts themselves to organic (or other) life found beyond the Earth.

The Moon Treaty (Art. VI-2) further recognises the right of states and their agents to 'collect on and remove from the Moon [and Mars] samples of its mineral and other substances' – which of course can include living matter as well, while the same Article

[11] Ibid 6.

[12] National Research Council of the National Academies, *Preventing the Forward Contamination of Mars* (The National Academic Press 2006) 13.

[13] United Nations Office for Outer Space Affairs, *International Space Law: United Nations Instruments* (United Nations 2017) 14.

[14] Ibid 32. Emphasis added.

[15] Ibid 5.

also allows for states and their agents ‘in the course of scientific investigations [to] also use mineral and other substances of the Moon in quantities appropriate for the support of their missions’.[16] It is an open question whether mining of resources for commercial ‘missions’ is also included here, but the refining of fuel, oxygen, and water on Mars for human settlements seems covered by this stipulation. Whether this is open to potential abuse is unknown, but given the actions by certain states on Earth—continued whaling for example—under the cover of scientific research, some doubt exists regarding whether states and their agents will always use only those resources truly needed for their missions, and not for commercial exploitation, keeping the non-appropriation principle in mind. Here, for example, national legalisation such as the Space Resource Exploration and Utilization Act of 2015 of the United States and the Act on the Exploration and Use of Space Resources of Luxembourg already clearly show the intentions of some states and their agents to harvest space-based resources on a commercial level, before legal questions around what constitutes appropriation have been settled.

However, the clearest stipulation in defence of celestial environments is found in Art. VII-1 (MOON), which declares that ‘States Parties shall take measures to prevent the disruption of the existing balance of its environment [the Moon and other celestial bodies, and in this case Mars], whether by introducing adverse changes in that environment, by its harmful contamination through the introduction of extra-environmental matter or otherwise. States Parties shall also take measures to avoid harmfully affecting the environment of the Earth through the introduction of extraterrestrial matter or otherwise’.[17] This non-contamination principle clearly prioritises the protection of both the environments of the Earth and of other celestial bodies from harmful contamination. While the Moon Treaty has relatively little international support, it entered into force in 1984 and states and their agents are thus behove to consider its stipulations and principles in their actions, even if they are not formal signatories.

Very recently (13 October 2020), eight states[18] signed the US-led Artemis Accords, which are intended to ‘increase the safety of [space] operations, reduce uncertainty, and promote the sustainable and beneficial use of space for all humankind’, and while many of the stipulations of the core outer space treaties are reaffirmed (such as registration of spacecraft and other objects), for the first time in an international accord stipulations are introduced for protecting heritage in outer space: ‘The Signatories intend to preserve outer space heritage, which they consider to comprise historically significant human or robotic landing sites, artifacts, spacecraft, and other evidence of activity on celestial bodies in accordance with mutually developed standards and practices’ (Section IX-1) and ‘The Signatories intend to use their experience under the Accords to contribute to multilateral efforts to further develop international practices and rules applicable to preserving

[16]Ibid 33.

[17]Ibid.

[18]Australia, Canada, Italy, Japan, Luxembourg, United Arab Emirates, United Kingdom, and the United States of America.

outer space heritage' (Section IX-2).[19] This is a very positive step in line with the topic of protecting heritage sites on the Moon, as discussed elsewhere.[20] The Artemis Accords also support 'safe and sustainable space activities' but, keeping the point made earlier in mind regarding commercial exploitation, also stipulate that 'The Signatories affirm that the extraction of space resources does not inherently constitute national appropriation under Article II of the Outer Space Treaty, and that contracts and other legal instruments relating to space resources should be consistent with that Treaty'.[21]

While it is positive that the Artemis Accords reinforce the primacy of the OST and other UN space treaties, the line between the extraction of space resources and appropriation is very blurry and open to dispute. Where does the boundary lie and when do extracted resources become appropriated? To answer this question, for the purposes of this chapter, the standard measure of appropriation will be proposed as follows: if resources are extracted and then sold (and hence property rights are conferred upon those resources) they are, de facto, appropriated.[22] For the purposes of the present inquiry, the concern of resource extraction is far less pronounced in the case of bodies which are *certainly* devoid of life as we know it, such as asteroids (keeping in mind that even the Moon may have harboured life at some point[23]), but what of cases where life exists? Shall we humans, as Voltaire cautioned, claim that all resources and objects in space exist solely for our own use, or shall we make exceptions where life is found? This question becomes even more pronounced if biospheres, however simple, exist on planetary scales such as (potentially) on Mars, and individuals with

[19]National Aeronautics and Space Administration, 'The Artemis Accords: Principles for Cooperation in the Civil Exploration and Use of the Moon, Mars, Comets, and Asteroids for Peaceful Purposes' (2020). https://www.nasa.gov/specials/artemis-accords/img/Artemis-Accords-signed-13Oct2020.pdf. Accessed 15 October 2020.

[20]André Siebrits, 'The Moon that Owns Itself: Exploring New Legal Avenues to Protect Cultural and Natural Heritage in Space' in Annette Froehlich (ed), *Protection of Cultural Heritage Sites on the Moon* (Springer 2020).

[21]National Aeronautics and Space Administration, 'The Artemis Accords: Principles for Cooperation in the Civil Exploration and Use of the Moon, Mars, Comets, and Asteroids for Peaceful Purposes' (2020). https://www.nasa.gov/specials/artemis-accords/img/Artemis-Accords-signed-13Oct2020.pdf. Accessed 15 October 2020.

[22]Using this standard, it is clear that the moon rocks and samples gathered by the Apollo missions were *not* appropriated, since most of the samples remain in secure storage with NASA and a few others were given away as gifts to foreign dignitaries, and even though some samples have gone missing and may have been sold to other parties, the intent behind gathering the Apollo samples was not to sell them (the same can be said of the Soviet samples automatically returned from the Moon), see: Mark Bosworth, 'What has happened to Nasa's missing Moon rocks?' (*BBC*, 20 February 2012). https://www.bbc.com/news/magazine-16909592. Accessed 19 October 2020.

[23]Kevin Loria, 'The moon may have once been able to support life—but learning more will require "an aggressive future programme of lunar exploration"' (*Business Insider*, 25 July 2018). https://www.businessinsider.co.za/earths-moon-could-have-supported-life-2018-7?r=US&IR=T. Accessed 15 October 2020.

great agency in the space arena have already openly called for altering the environment of Mars to better suit human needs (Elon Musk's proposal to detonate nuclear bombs over the Martian poles comes to mind here[24]).

10.3 Life on Mars

> The surface area of Mars is exactly as large as the land area of the Earth. A thorough reconnaissance will clearly occupy us for centuries. But there will be a time when Mars is all explored; a time after robot aircraft have mapped it from aloft, a time after rovers have combed the surface, a time after samples have been returned safely to Earth, a time after human beings have walked the sands of Mars. What then? What shall we do with Mars?
>
> Carl Sagan, *Cosmos* (Ballantine Books 2013) 138.

Since ancient times, humans have been preoccupied by the idea of finding life on other worlds. The earliest such ideas revolved around conceptions of cosmic pluralism, in which there were many inhabited worlds, long before the ideas of the modern scientific revolution came to be, and which revived the idea of a plurality of worlds. In what is hailed as the first work of science fiction, the author Lucian of Samosata wrote in his book *A True Story* accounts of voyages in outer space where he 'brokers war and peace between the inhabitants of the sun and moon'.[25] Humanity has thus been interested in the search for life beyond the Earth for most of our history, which in turn begs the question, to paraphrase Carl Sagan, what shall we do with Mars if we find life there?

Despite some claims that humanity's first encounter with life beyond the Earth will be with intelligent beings,[26] given the rarity of such beings (those at least being capable of advanced reasoning, scientific thought, and communication) in relation to simpler life on the Earth (one species out of almost nine million currently, over a period of 3.5 billion years), as well as the recent discovery of phosphine gas on Venus, coupled with no confirmed sign of intelligent life hitherto on Mars (or anywhere else in space), the present inquiry will be limited to simpler forms of life. However, this does not diminish the importance of the question or of its consequences.

Up to the present (October, 2020), Mars has had 26 successful visits from the Earth, in the form of four flybys (Mariners 4, 6, and 7, and MarCO), 12 orbiters (Mars 2 and 3, Mariner 9, Viking 1 and 2 orbiters, Mars Global Surveyor, Mars Odyssey, Mars Express, Mars Reconnaissance Orbiter, Mars Orbiter Mission, MAVEN, and

---

[24] Mike Wall, 'Elon Musk Floats 'Nuke Mars' Idea Again (He Has T-Shirts)' (*Space.com*, 17 August 2019). https://www.space.com/elon-musk-nuke-mars-terraforming.html. Accessed 22 October 2020.

[25] Lucian of Samosata, 'A True Story' (*Sacred Texts*, no date). https://www.sacred-texts.com/cla/luc/true/index.htm. Accessed 19 October 2020.

[26] Eric Mack, 'The First Alien Life We Find Is More Likely To Be Intelligent Than Not, SETI Expert Says' (*Forbes*, 19 February 2020). https://www.forbes.com/sites/ericmack/2020/02/19/why-the-first-alien-life-we-find-is-more-likely-to-be-intelligent-than-not/#334a68253f10. Accessed 19 October 2020.

ExoMars Trace Gas Orbiter), six landers (Mars 3, Viking 1 and 2, Mars Pathfinder, Phoenix, InSight), and four rovers (Sojourner, Spirit, Opportunity, and Curiosity). These exclude failures or partial failures, and gravity assists to other destinations. These missions belonged (and continue to belong under international law) to the United States, Soviet Union/Russia, European Space Agency, and India. A further three missions are underway to Mars, by China (Tianwen-1), the United Arab Emirates (Emirates Mars Mission), and the United States (Perseverance and Ingenuity).

What do these missions say about the prospects of finding life currently on Mars? Perhaps the first discovery of signs of life came from the 1976 Viking lander Labeled Release (LR) life detection experiment, which returned 'a total of four positive results, supported by five varied controls, streamed down from the twin Viking spacecraft landed some 4000 miles apart' and the 'data curves signaled the detection of microbial respiration on the Red Planet'.[27] However, the results were dismissed when the accompanying Viking Molecular Analysis Experiment 'failed to detect organic matter', leading NASA to determine that the 'LR had found a substance mimicking life, but not life'.[28] Since the experiment was never repeated, and subsequent investigations focused on finding evidence for past life, and the conditions for past life, on Mars, it remains an open question what exactly the LR found. Other studies have found chemical compounds called 'thiophenes' in 3.5 billion-year-old Martian soil, which 'could, potentially, be organic' or which could have come from asteroids.[29] Transient methane plumes, including those detected by Curiosity, with a seasonal variation, could also be potentially of organic origin, although other explanations do exist.[30] Underground reserves of liquid water, although very salty or briny, have also been detected on Mars.[31] There is also the controversial Allan Hills 84001 Martian meteorite analysed in 1996 which potentially suggested the presence of ancient life on Mars.[32] While these tantalising clues are not conclusive proof of life on Mars, they do prompt us as human explorers to take seriously the question of the legal, moral, and ethical dilemmas a confirmed discovery of even the simplest living organisms

[27] Gilbert V. Levin, 'I'm Convinced We Found Evidence of Life on Mars in the 1970s' (*Scientific American*, 10 October 2019). https://blogs.scientificamerican.com/observations/im-convinced-we-found-evidence-of-life-on-mars-in-the-1970s/. 19 October 2020.

[28] Ibid.

[29] Tristan Greene, 'Scientists discover strong evidence of life on Mars' (*TNW*, 7 March 2020). https://thenextweb.com/science/2020/03/07/scientists-discover-strong-evidence-of-life-on-mars/. Accessed 19 October 2020.

[30] Andrew Good and Alana Johnson, 'Curiosity's Mars Methane Mystery Continues' (*NASA*, 23 June 2019). https://www.nasa.gov/feature/jpl/curiosity-detects-unusually-high-methane-levels. Accessed 19 October 2020.

[31] Jonathan O'Callaghan, 'Water on Mars: discovery of three buried lakes intrigues scientists' (*Nature*, 28 September 2020). https://www.nature.com/articles/d41586-020-02751-1. Accessed 19 October 2020.

[32] Charles Q. Choi, 'Mars Life? 20 Years Later, Debate over Meteorite Continues' (*Space.com*, 10 August 2016). https://www.space.com/33690-allen-hills-mars-meteorite-alien-life-20-years.html. Accessed 19 October 2020.

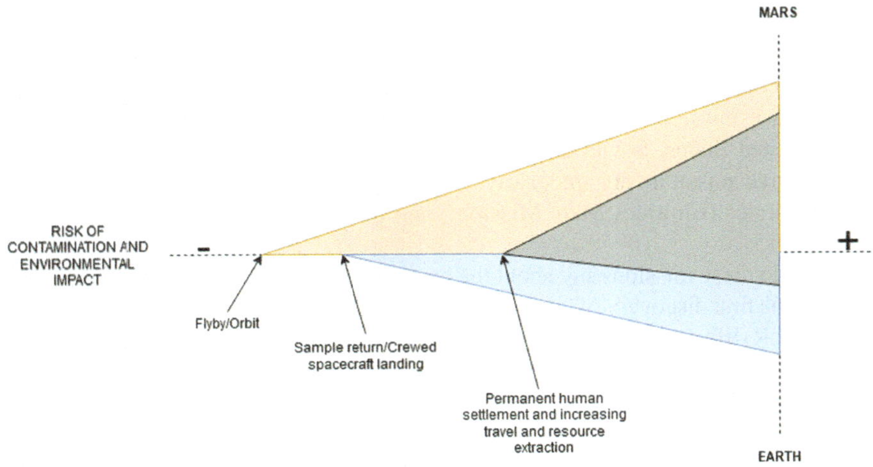

Fig. 10.2 Earth-Mars cross-contamination model in the context of increasing human activities

would present. The present time (early 2020s) offers the best opportunity for grappling with these issues, before the human presence on Mars becomes established and permanent. We are thus currently living in a narrow window of time at a 'historical branch point' and '[w]hat we do with our world [and with Mars, soon to be our world as well if current plans come to fruition] in this time will propagate down through the centuries and powerfully determine the destiny of our descendants and their fate, if any, among the stars'.[33]

One of the core considerations related to life on Mars is the abovementioned stipulation on non-contamination. While the focus of this chapter is on the protection of extraterrestrial environments, specifically Mars, Fig. 10.2 highlights the potential bi-directional adverse effects through contamination and environmental disruption and degradation as human activities and contacts grow between Earth and Mars. Thus far, human impacts on the Martian environment have been minimal, with the least threat posed by flybys or orbiting craft around Mars, and relatively little by landers and rovers, even though there is always a potential risk of contamination despite the best efforts made by spacecraft manufacturers. For example, the Israeli Beresheet probe which crashed on the Moon carried tardigrades as part of its cargo, which scientists believe are 'alive and well'.[34] Such reckless contamination of celestial bodies notwithstanding, an experiment conducted on the International Space Station has shown that bacteria exposed to the vacuum of space can survive 'for years'.[35]

[33]Carl Sagan, *Cosmos* (Ballantine Books 2013) 228.

[34]Ian Sample, 'Tardigrades may have survived spacecraft crashing on moon' (*The Guardian*, 6 August 2019). https://www.theguardian.com/science/2019/aug/06/tardigrades-may-have-survived-spacecraft-crashing-on-moon. Accessed 19 October 2020.

[35]Max G. Levy, 'Scientists Discover Exposed Bacteria Can Survive in Space for Years' (*Smithsonian Magazine*, 26 August 2020). https://www.smithsonianmag.com/science-nature/scientists-discover-exposed-bacteria-can-survive-space-years-180975660/. Accessed 19 October 2020.

There is thus a continuing threat of human contamination of Mars as missions to the planet (and especially human visits) increase. Conversely, the first threat to Earth emerges when samples are returned to the Earth for analysis, and when humans, having been in extended contact with the Martian environment, return to Earth (back contamination). Of course, space agencies follow strict protocols to avoid any risk to the Earth, such as the quarantine of the Apollo astronauts in the case of the Moon, but the risk to the Martian environment rapidly grows as humans live and work on the planet and extract resources there. As in the case of the Moon, human waste will also be left behind,[36] and a permanent settlement will drastically increase the risk to Mars and make large-scale contamination practically inevitable and beyond the control of space agencies and their governments back on Earth.

As the model in Fig. 10.2 suggests, the primary contamination impact of human activities will be on Mars, not Earth. Even with the hitherto unconfirmed presence of life on Mars, the Mars Sample-Return mission, planned to bring back samples gathered by Perseverance, will undoubtedly 'have a containment procedure that is stupefyingly reliable' given the risks.[37] While great care and consideration is given to prevent the contamination of Mars and other worlds by human spacecraft,[38] failures do happen as the Beresheet probe showed. Another concern is whether international space law is keeping up with the dangers, alongside greater moral and ethical considerations. One of the ways in which the law can adapt to face these challenges is by considering the application of the growing trend mentioned earlier to grant natural features and biomes an independent legal status.

10.4 Lessons from the Moon that Owns Itself

> What shall we do with Mars? There are so many examples of human misuse of the Earth that even phrasing this question chills me. If there is life on Mars, I believe we should do nothing with Mars. Mars then belongs to the Martians, even if the Martians are only microbes. The existence of an independent biology on a nearby planet is a treasure beyond assessing, and the preservation of that life must, I think, supersede any other possible use of Mars.
>
> Carl Sagan, *Cosmos* (Ballantine Books 2003) 138.

---

[36] See for example: Hugo Lopez, 'The Protection of Cultural Heritage Sites on the Moon: The Poo Bags Paradox' in Annette Froehlich (ed), *Protection of Cultural Heritage Sites on the Moon* (Springer 2020).

[37] Carl Sagan, *Cosmos* (Ballantine Books 2013) 137.

[38] See: Margaret S. Race, James E. Johnson, James A. Spry, Bette Siegel, and Catharine A. Conley (eds), 'Planetary Protection Knowledge Gaps for Human Extraterrestrial Missions: Workshop Report' (*NASA*, 24–26 March 2015). https://strives-uploads-prod.s3.us-gov-west-1.amazonaws.com/20160012793/20160012793.pdf?AWSAccessKeyId=AKIASEVSKC45ZTTM42XZ&Expires=1603114859&Signature=E%2BDHL6f4RMVcyK%2Fr8S58uTleFAo%3D. Accessed 19 October 2020; National Research Council of the National Academies, *Preventing the Forward Contamination of Mars* (The National Academic Press 2006).

In recent years, and in response to the natural environment which has been increasingly 'polluted, mined, diverted and degraded to a shocking extent', several countries have started taking innovative new steps to protect vital and significant areas by granting (for example to the Ganges and Yamuna rivers in India) the 'status of a legal person, with all corresponding rights, duties and liabilities… in order to preserve and conserve them'.[39] Other countries have done the same, such as New Zealand in the case of the Whanganui River and ecosystem, and the aforementioned Colombia in terms of the Amazon. By recognising, for example, 'the intrinsic rights of rivers, beyond their use for humans', these states and their legal systems have taken a bold step towards ending the wanton and often unpunished abuse of the environment, with the understanding that '[r]ecognising such rights does not stop all human use of nature, but means that our actions must not interfere with the ability of ecosystems and species to thrive'.[40]

In a previous publication,[41] the argument was put forth that this growing list of cases should be studied for the potential application of this model to celestial bodies in outer space. Here, the argument is made specifically for Mars—and especially if life is confirmed to exist there in any form. While it is granted that such an idea will have its opponents, especially those who wish to exploit natural resources in space for private commercial gain (leading to appropriation given the earlier definition provided and thus violating the OST) and well as those who wish to extend national and military ambitions into space (for example the Astropolitik policy espoused by Dolman,[42] who argues that 'by using its current and near-term capacities, the United States should endeavor at once to seize military control of low-Earth orbit',[43] and who happens to be a Professor of Strategy at the US Air Force Air Command and Staff College (ACSC)), it nevertheless offers the best approach to preserve the spirit of the OST and other treaties in the face of new challenges.

Since it is already accepted through the legal precedent of the five outer space treaties that Mars is beyond the national sovereignty of states on Earth, and beyond *de jure* appropriation or harmful interference by mankind, and it is not a great leap to assert that the planet, or parts thereof where life might be found, exist in their own legal right. As on Earth, this does not preclude human access to, or use of, Mars and its natural resources, including potential living organisms there, but it does mean that far greater care would have to be taken to avoid infringing on the legal rights of the planet or its pre-existing inhabitants and their ability to survive and thrive. By possessing

---

[39] Ashish Kothari, Mari Margil, and Shrishtee Bajpai, 'Now rivers have the same legal status as people, we must uphold their rights' (The Guardian, 21 April 2017). https://www.theguardian.com/global-development-professionals-network/2017/apr/21/rivers-legal-human-rights-ganges-whanganui. Accessed 20 October 2020.

[40] Ibid.

[41] André Siebrits, 'The Moon that Owns Itself: Exploring New Legal Avenues to Protect Cultural and Natural Heritage in Space' in Annette Froehlich (ed), *Protection of Cultural Heritage Sites on the Moon* (Springer 2020).

[42] Everett C. Dolman, *Astropolitik: Classical Geopolitics in the Space Age* (Frank Cass 2002) 156–165.

[43] Ibid 157.

legal personhood, there would also be a pathway to prosecute violators of Martian legal rights. The only court that could establish these rights would be the International Court of Justice (ICJ), to which all United Nations members are party and which is the only body which can adjudicate international disputes. Of course, since advisory proceedings have to be initiated by UN agencies and organs, the task would have to fall on the United Nations Office for Outer Space Affairs (UNOOSA), through the auspices of the United Nations Committee on the Peaceful Uses of Outer Space (UNCOPUOS). This is not to say major and possibly insurmountable challenges would not be faced by such a proposal, especially given the difficulties in reaching agreement on major issues within UNCOPUOS, which has led the committee to favour the adoption of non-binding measures instead of major legally-binding treaties in recent times. However, there is no reason this proposal cannot also be a non-binding one at first, eventually leading to a new norm and customary law.

As such, the argument here is that the committee, and the legal and space communities, will need alternatives in dealing with mounting challenges to the outer space regime in the coming years. This will be especially true in the case of finding life on Mars (or Venus or other celestial bodies like Europa). By taking up the cause of furthering the legal status of Mars and other celestial bodies and thus strengthening and expanding the corpus of space law instead of casting it aside as some propose, this chapter is part of what will become a growing debate in the next few years. In agreement with Carl Sagan's position, if life is found on Mars, its interests should supersede our (alien) interests in exploiting and altering the planet for our own purposes. Numerous examples on Earth have demonstrated, and continue to demonstrate, how narrow commercial or ideological interests can have devastating consequences for the natural environment on which we all depend. We cannot make the same mistakes on other worlds and establish a predatory precedent for our descendants. By recognising that celestial bodies and their inhabitants have legal rights akin to those of the Colombian Amazon, the Ganges, and other natural bodies and features, competition between our interests and those of pre-existing life on Mars can be avoided, or at the very least minimised. After all, Art. VII-1 (MOON) makes it clear that humans should not alter or disrupt the existing environmental balance on celestial bodies, even in the absence of life (thus halting Elon Musk's terraforming plans despite the US not being a signatory of the Moon Treaty). How much more critical then is preserving and protecting these environments in the presence of life?

10.5 Conclusion

> If microorganisms were to be found in one of these special regions, or anywhere on Mars, it would be a huge deal. But the very possibility raises the important issue of keeping Earthling hitchhikers out. We know how problematic (and in some cases, disastrous) invasive species on Earth have been: What would happen to life on Mars if it's overwhelmed or outcompeted by microbes from Earth before we even have a chance to study it?

Starre Vartan, 'When We Go to Mars, How Will We Protect the Microbes From Us?' (*Mental Floss*, 27 November 2016). https://www.mentalfloss.com/article/87459/when-we-go-mars-how-will-we-protect-martian-life-us. Accessed 22 October 2020.

In the introduction of this chapter, the question posed by Carl Sagan was echoed—who speaks for Earth? His answer was clear—it is not the superpowers, it is us, all of us, who inhabit the Earth. This chapter has posed the same question for the Red Planet—who speaks for Mars? Since there is no sign of intelligent life capable of communicating with humans anywhere in the known solar system, it is once again we who must speak for Mars. While some may dismiss the question of how we should treat even the simplest life on other worlds as trivial or irrelevant, this very question will define the future of the human species in outer space. If simple life is plentiful, as some argue,[44] and as some clues from Venus, Mars, and even Enceladus have suggested, this is a question we must answer soon. The best way of doing this is through existing space law, which embodies many of the noblest human values, and which argues that all humans should share in, and benefit from, the great saga of outer space exploration, and that it should be a realm of peace and non-appropriation. Those in the legal and space communities who believe in, and support, this vision of our future in outer space should not allow themselves to be naive enough to think that opponents to this regime do not exist, or that increasing human activities in space will not provoke new legal questions and challenges that may unravel the outer space regime.[45]

For this reason, this chapter has considered the example of R2P for outer space (this topic will be further explored in a future study) to reflect on the responsibilities of humans in dealing with celestial bodies, especially if they harbour life. The example of illustrative cases on Earth in finding new ways of regulating our dealings with the natural environment was proffered as a future model for dealing with the environments of celestial bodies beyond the Earth as well, especially because while some space agencies may have detailed and strict planetary protection procedures, new commercial and national space actors are entering the space arena ever year, and how confident can we be that their actions and approaches will always be equally rigorous? Therefore, at this critical historical junction, on the precipice of sending human explorers to Mars and other worlds, immense legal, moral, and ethical questions—and challenges—loom. However, there is no reason the solutions cannot also be equally immense. Moreover, the relationship between the two worlds of Mars and Earth is dialectical in nature, and as we learn more about Mars and develop our relationship with that planet, we will also learn more about the Earth and the comparative rarity and value of life here. Even without life existing on Mars or elsewhere beyond the Earth, our responses to the questions posed by the mere possibility of the

[44]'Study: Simple Life Forms are Common throughout Universe' (*Sci-News.com*, 19 December 2017). http://www.sci-news.com/paleontology/simple-life-forms-common-universe-05549.html. Accessed 22 October 2020.

[45]These issues are discussed in greater depth in a forthcoming publication: André Siebrits, 'Hegemonic Struggle in the Global Space Arena: Can the Centre Hold?' in Annette Froehlich (ed), *Outer Space and Popular Culture: Influences and Interrelations, Volume 2* (Springer, forthcoming).

existence of such life will teach us much about ourselves and the kind of species we want to be. Perhaps this will be the greatest, and final, gift of Mars.

André Siebrits is a South African researcher focusing on the space arena (especially in developing world contexts), as well as education and the use of educational technologies, and international relations (particularly in the Global South). He is currently working with the European Space Policy Institute (Vienna), and has experience as an e-learning researcher and as an African political risk analyst. He graduated with a Master of Arts in International Studies from the University of Stellenbosch, where his research revolved around theories of international relations. He is also an Associate Editor of the Springer series Southern Space Studies. He is currently a PhD Candidate at the Department of Political Studies at the University of Cape Town (UCT), where his research focuses on the role of the Global South in the space arena, especially in relation to governance, seen from an international relations perspective. Mr. Siebrits has several publications in the e-learning and space fields and has written on the space-education ecosystem for sustainability and the role of educational technologies in Africa, on integrated space for African society (legal and policy implementation of space in African countries—specifically Algeria, Morocco, Tunisia, and Zimbabwe), and on the African space arena. He has also lectured at the UCT SpaceLab (Space and Society course) on the African space arena and on the role of educational technologies in space education in Africa.

Chapter 11
Crewed Space Mission Waste-Streams and Impact on Human Exploration of Mars

Samuel Anih

Abstract Crewed space missions have historically used and still employ basic waste-stream management strategies to dispose of mission waste-streams. Previous space waste management methods include ejecting waste into space, dumping waste on the surface celestial body, storing them in spacecraft for return to Earth or incineration during re-entry. Currently on board the International Space Station, waste-streams are loaded onto cargo vehicles and incinerated in the Earth's atmosphere, a luxury future long-duration crewed mission away from Earth will not have. With the prospect of long-duration crewed spaceflights to Mars in the 2030s, new strategies have to be developed to address the challenges of waste-stream management during long-duration human spaceflight missions, most especially on the Martian surface. This is important for the fidelity of exploration goals which include various forms of science that will target the search for possibility of extinct and extant life among others. This article explores the possible impact of waste-streams on exploration of Mars when humans eventually set foot on the planet, potentially bringing various waste-streams to an erstwhile pristine celestial body. The impact of waste-streams on crewed space missions to Mars and possible approaches to waste-borne contamination mitigation are also discussed.

11.1 Introduction

From 1968 to 1972, twenty-four humans left the Earth's orbit and flew around the Moon on nine missions. Out of these, only twelve walked on the Moon surface and since then humans have not yet ventured beyond low-Earth orbit. Human spaceflight in the past four decades has focused on learning to live and work in space in a variety of spacecraft including stations, such as Salyut, Skylab, Mir and currently the

S. Anih (✉)
SpaceLab, Department of Electrical Engineering, University of Cape Town, Cape Town, South Africa
e-mail: anihsammy@gmail.com

OAU, ARCSSTE-E, OAU Campus, Ile-Ife, Nigeria

A. Froehlich (ed.), *Assessing a Mars Agreement Including Human Settlements*, Studies in Space Policy 30, https://doi.org/10.1007/978-3-030-65013-1_11

International Space Station (ISS). The delay in extending exploration to more distant targets in the solar system has partly to do not only with cost and lack of political will, but also largely to do with developing the necessary technical capabilities to sustain human life for long periods of time in deep space. There is currently a resurgence of interest in human space exploration of the Moon and Mars, with some of the more optimistic scenarios predicting human landings on Mars before 2030.

There are material and technological costs involved in putting humans in space since the advent of human spaceflight and these also come with the challenges of keeping the crew alive and comfortable during those missions. The space environment and what is known of other planetary surfaces in the solar system show stark differences from the environmental conditions on the surface of the Earth. Basic needs for human survival such as breathable air, water, food and adequate waste disposal system are naturally handled by Earth's ecological system, but in space these natural processes have to be mimicked through the use of mechanical and physicochemical equipment.[1] After Yuri Gagarin's historic flight, subsequent human spaceflight longer than two hours in duration have to contend with waste management issues in one way or the other. The Gemini 7 mission, in 1965 which lasted almost 14 days in low Earth orbit inside cramped quarters of the Gemini capsule necessitating austere waste collection strategies and stowage of waste streams including human waste, described by the crew as 'distasteful'.[2] Subsequent crewed missions have adopted various waste management strategies as crew members began have more extended missions in space.

11.1.1 Human Spaceflight Waste Management Experience

Waste-streams from crewed space missions are usually generated from consumables supplied from Earth, such as waste from food preparation, payload related operations, crew metabolic output and subsystem operation.[3] Solid wastes from spaceflight comprise paper, inedible plant biomass (IPB), plastic, cardboard, faeces urine concentrates, waste water concentrates, etc.[4] The waste is first collected, separated according to composition if needed, compressed, stabilized and then stored for disposal. Human waste is placed inside metal containers, while housekeeping

[1]Carol Norberg, *Human Spaceflight and Exploration* (Springer Praxis Books, Springer-Verlag Berlin Heidelberg 2013).

[2]James Lee Broyan Jr., 'Waste Collector System Technology Comparisons for Constellation Applications' [2007] International Conference On Environmental Systems. https://ntrs.nasa.gov/archive/nasa/casi.ntrs.nasa.gov/20070016696.pdf.

[3]Ernst Messerschmid and Reinhold Bertrand, *Space Stations Systems and Utilization* (Springer Verlag Berlin Heidelberg 1999).

[4]Michael A. Serio, Joseph E. Cosgrove and Marek A. Wójtowicz, 'Use of Pyrolysis Processing for Trash to Supply Gas (TtSG)' [2014] International Conference on Environmental Systems 1.

and crew provision wastes are stored in plastic bags.[5] Some of the wastes do have biological origin and require special handling.

Proper waste-stream management is an integral part of any crewed mission to space, most especially for missions lasting more than a few hours. Almost all the crewed space missions so far have adopted a similar approach to waste-stream management consisting of one-time-use of supplies. The resultant waste-streams are manually compacted and consolidated or stabilized and then stowed for final disposal.[6] There is a significant house-keeping overhead of waste-stream management in relation to logistics during crewed space missions[7] and this has a direct effect on crew operations, which leads to waste-stream management being described as one of the functions that require significant amounts of crew time[8]—a precious resource during crewed spaceflight. NASA's *Human Integration Design Handbook*[9] and *Non-Recoverable Cargo (Trash/Waste) Management Plan*[10] emphasize the design and goal of trash management with regards to minimization, odour and contamination control as well as the containment of hazardous trash.

The International Space Station, a human-made laboratory maintained in low Earth orbit (LEO) at an altitude of 330–435 km, has the advantage of being permanently close to the Earth. Rudimentary waste management strategies were adopted over the years for waste-streams generated inside spacecraft. Currently on the ISS, waste-streams are stowed for months (Fig. 11.1), removed regularly and incinerated using the un-controlled re-entry of expendable cargo vehicles in the Earth's atmosphere.

Waste-stream in the context of this article refers to various forms of waste packets from crewed space missions; they can be from biotic and non-biotic sources. Waste-streams include food packaging, paper, tape, soiled clothing, metabolic

[5]Molly S. Anderson, Michael K. Ewert and John F. Keener, 'Advanced Life Support Baseline Values and Assumptions Document'. https://ntrs.nasa.gov/archive/nasa/casi.ntrs.nasa.gov/20180001338.pdf.

[6]Kathryn Hurlbert and others, 'Draft Human Health, Life Support and Habitation Systems Technology Area 06'. https://www.nasa.gov/pdf/500436main_TA06-HHLSHS-DRAFT-Nov2010-A.pdf; Masamichi Yamashita and Raymond M. Wheeler, 'Habitation in Space' in Malcolm Macdonald and Viorel Badescu (eds), *The International Handbook of Space Technology* (Springer-Verlag Berlin Heidelberg 2014).

[7]Peter H. Needham and Peter H. Needham, 'Spacecraft Housekeeping and Its Relationships to Logistics' (1971) 1. http://commons.erau.edu/space-congress-proceedings/proceedings-1971-8th/session-8/3.

[8]Waste management is described as one of the life support functions that could critically impact crew time in section 3.2.7 of the *Life Support Baseline Values and Assumptions Document* Anderson, Ewert and Keener (n 6).

[9]NASA, 'Human Integration Design Handbook' (2010) SP-2010-3407. https://ston.jsc.nasa.gov/collections/TRS/_techrep/SP-2010-3407.pdf.

[10]NASA, 'Non-Recoverable Cargo (Trash/Waste) Management Plan International Space Station Program Baseline' 1. http://www.spaceref.com/iss/ops/SSP50481.nonrec.cargo.plan.pdf.

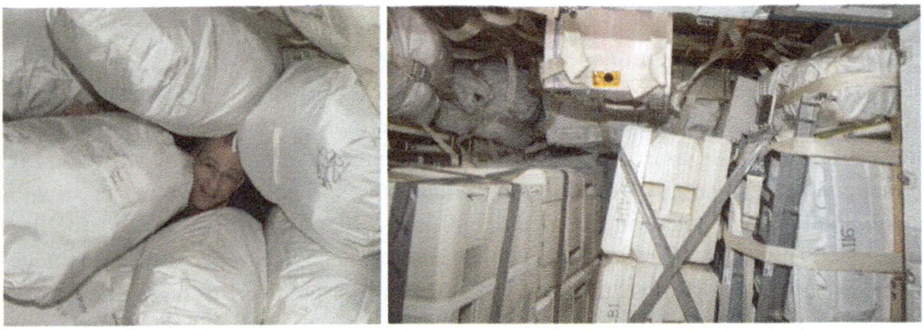

Fig. 11.1 Waste stowage on ISS (left), appreciable volume on the station is occupied by stowed waste/trash (right) until they are removed by cargo vehicles (NASA, 'Expedition 30 Flight Engineer Don Pettit' (2012). https://www.nasa.gov/mission_pages/station/multimedia/gallery/iss030e020706.html. Accessed 10 July 2019)

Table 11.1 Types and sources of wastes in the crewed space environment

Waste type	Waste sources
Liquid, biological, decomposable	Hygiene water, metabolic water, respiration/transpiration water, urine, liquid faeces
Solid, biological, decomposable	Solid faeces, waste with bound water, solids from urine/sweat/hygiene water, clothes
Gaseous, metabolic	CO_2, trace gases, methane
Liquid, nonrecoverable	Medicines, payload/experiment products or effluent
Solid, nonrecoverable	Spare parts, plastics, metals

Susan Doll, 'Environmental Control and Life Support Systems (ECLSS)' in Wiley J Larson and Linda K Pranke (eds), *Human Spaceflight: Mission Analysis and Design* (McGraw-Hill 1999)

waste, expended hygiene supplies, inedible biomass and failed equipment parts.[11] A breakdown of waste sources in the spaceflight environment is shown in Table 11.1.

11.2 Future Long-Duration Crewed Space Missions Waste Management

The question of waste management has never had to be solved in a holistic manner throughout the history of human space exploration because missions have never been long enough, or far enough away from Earth, that waste management has become a mission-critical issue.

[11] Donald Rapp, 'Mars Life Support Systems' [2006] The Mars Journal 72. http://www.marsjournal.org/contents/2006/0005/.

Human spaceflight to destinations further away from the Earth-Moon system imposes additional challenges for proper waste management due to several constraints, such as limited resources, absence of constant resupply of consumables, limited habitable volume inside the spacecraft, isolation from the Earth system and waste stowage for a longer period of the mission.[12] In such cases the ISS mode of waste management would no longer be a suitable option as waste-streams generated would have to be managed within the spacecraft during the transit phase and on the surface when the astronauts finally set foot on Mars.

Crewed space missions to other celestial bodies would require a proper understanding of environmental control and life support systems (ECLSS) to keep the astronauts alive while at the same time preserving consumables that would be necessary to support such missions. Consequently, the need to adequately manage the waste-streams generated from such missions would also arise with increase in waste volume within the spacecraft or habitat. The mass and volume of consumables including the resultant waste generated directly impacts the launch mass of the launch vehicle as well as various manouevres needed during the course of the crewed mission respectively.

11.3 Mars as a Destination for Crewed Space Missions

Various destinations and scenarios have been suggested for future long-duration crewed space missions but there is presently no agreement amongst experts on the priorities.[13] Some advocate returning to the Moon first as a stepping stone to Mars. Others suggest that a "Mars direct"[14] strategy is the best way forward.[15,16] Part of the reason for modification of these destinations have been the evolution of various architectures[17] and constant policy changes with attendant budgetary cuts from various administrations.

[12]Harry W. Jones, Edward W. Hodgson and Mark H. Kliss, 'Life Support for Deep Space and Mars' [2014] 44th International Conference on Environmental Systems 1.

[13]Jean Marc Salotti and Richard Heidmann, 'Roadmap to a Human Mars Mission' (2014) 104 Acta Astronautica 558. http://dx.doi.org/10.1016/j.actaastro.2014.06.038.

[14]This Mars direct refers to a strategy of going to Mars without first returning to the Moon. It is different from the *Mars Direct* originally proposed in 1990 by Robert Zubrin and David Baker of the then Martin Marietta Corporation as a cost-effective human mission to Mars achievable with current technology.

[15]Robert M. Zubrin and David A. Baker, 'Mars Direct: Humans to the Red Planet by 1999' (1992) 26 Acta Astronautica 899.

[16]Jean Marc Salotti, 'Robust, Affordable, Semi-Direct Mars Mission' (2016) 127 Acta Astronautica 235. http://dx.doi.org/10.1016/j.actaastro.2016.06.004.

[17]NASA, 'NASA's Journey to Mars, Pioneering Next Steps in Space Exploration' [2015] NASA's journey to Mars 1. https://www.nasa.gov/sites/default/files/atoms/files/journey-to-mars-next-steps-20151008_508.pdf.

The nature of locations and destinations chosen for crewed space missions have implications on the operations and habitation of the astronauts – most importantly it will affect the volume of consumables needed and the waste management mode to be adopted. The farther the distance from Earth, the more comprehensive and robust the life support functions needed to support the crew have to be.

Mars as a potential crewed space mission destination has several characteristics and features that have made it suitable and promising for human exploration. This would require a tailored waste management approach.

Mars is by far the most explored planet in the solar system. Robotic orbiters and landers have explored the planet for data over the years. Rovers such as *Pathfinder, Spirit* and *Opportunity* and most recently *Curiosity* have provided in situ information about the planet.

The interest in Mars as a prime destination for human exploration stems from the following:

(i) The planet possesses all the essential ingredients required to sustain life and these ingredients could be extracted using in situ resource utilization (ISRU) during exploration.[18,19,20]
(ii) Results from robotic explorations of Mars show signs of a planet that was habitable in the distant past with more Earth-like conditions than it has today.[21] This encourages the search for extant or extinct life—or signatures of both.[22]
(iii) *Lessons learned from Mars history could help to predict the fate of Earth's future.*

Mars' features suggest the planet has the potential as a future settlement for humankind, with the capacity to provide a suitable place to extend the presence of the human species beyond Earth. This will provide "insurance" for humanity against being wiped out by a cataclysmic event like the one that wiped out the dinosaurs about 65 million years ago and help ensure the survival of the human race.

Unlike the Moon, which is 384,000 km away from Earth (three days travel time with current technology), Mars' closest approach to Earth can vary quite significantly (Fig. 11.2); the closest in recent times was in 2003 when it was 55.7 million kilometers. This not only affects the transit time, but also the overall requirements to

---

[18]National Academy of Sciences, 'Vision and Voyages for Planetary Science in the Decade 2013-2022'. http://www.nap.edu/catalog.php?record_id=13117.

[19]Robert W. Moses and Dennis M. Bushnell, 'Frontier In Situ Resource Utilization for Enabling Sustained Human Presence on Mars' [2016] NASA Scientific and Technical Information. https://ntrs.nasa.gov/search.jsp?R=20160005963%0Ahttps://ntrs.nasa.gov/archive/nasa/casi.ntrs.nasa.gov/20160005963.pdf.

[20]Morgan B. Abney, Jay L. Perry and Gerald B. Sanders, 'A Discussion of Integrated Life Support and In Situ Resource Utilization Architectures for Mars Surface Missions' [2018] 48th International Conference on Environmental Systems.

[21]Marlies Arnhof, 'Design of a Human Settlement on Mars Using In Situ Resources' [2016] 46th International Conference on Environmental Systems 1. http://hdl.handle.net/2346/67561.

[22]Bret G. Drake, Stephen J. Hoffman and David W. Beaty, 'Human Exploration of Mars, Design Reference Architecture 5.0' [2010] IEEEAC.

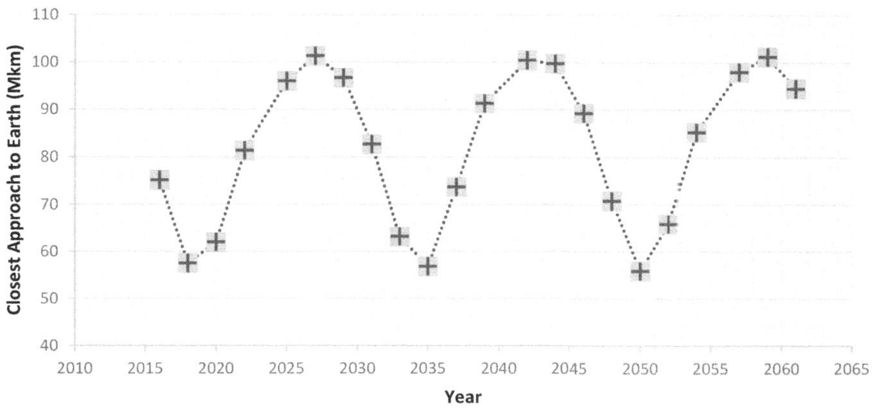

Fig. 11.2 Minimum distance between the orbits of Earth and Mars from 2016 to 2061, measured in millions of kilometres—data from Hartmut Frommert, 'Mars Opposition' (http://spider.seds.org, 2008)

sustain the crew during Mars crewed missions. The long-duration away from Earth also has overarching effect on overall mission waste management strategies along the various gravity wells transited throughout the mission and on the surface.

The NASA's *Mars Design Reference Architecture (DRA) 5.0*[23] is a mission reference document designed by NASA on how to get humans to Mars. It is the most representative of the crewed mission architectures to Mars to date, finalized in 2009 with an addendum in 2014. It describes sequence of missions and supporting infrastructure for a crewed mission to Mars with a crew of six. The choice of a crew of this size was adjudged to be good compromise between the skillset required by the astronauts and level of effort presented by missions due to the complexity and duration while balancing this with the magnitude of the systems and infrastructure needed to support the crew.[24]

Crewed missions to Mars consist of two distinct categories of mission profile: the opposition and conjunction class (Fig. 11.3), a naming convention which refers to the relative position of Mars in comparison to Earth and the Sun midway through the mission.[25] The opposition class mission has longer total transit times with shorter period in Mars vicinity, typically less than 90 days while conjunction class mission results in short transit times with longer stay on Mars surface.[26] The DRA 5.0, a conjunction class, long-duration stay (500 days or more on the surface), was chosen

[23] Bret G. Drake, 'Human Exploration of Mars Design Reference Architecture 5.0' (2009). https://www.nasa.gov/pdf/373665main_NASA-SP-2009-566.pdf.

[24] Bret G. Drake, 'Human Exploration of Mars : Challenges and Design Reference Architecture 5.0' (2010) 12 3578.

[25] John M. Colombi and others, 'Conceptual Design Using Executable Architectures for a Manned Mission to Mars' (2015) 9 IEEE Systems Journal 495.

[26] Mattfeld and others (n 29).

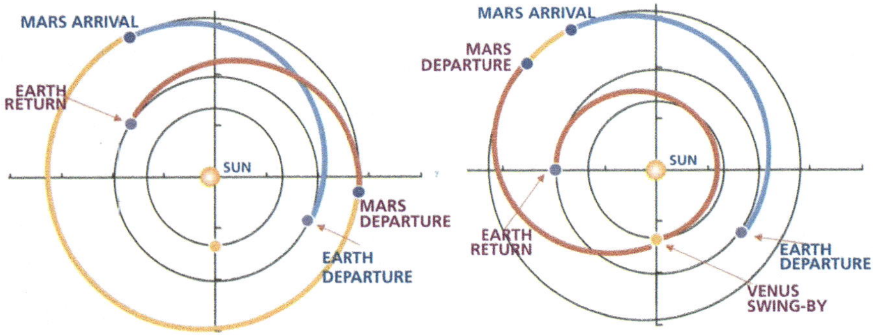

Fig. 11.3 Representative conjunction trajectory concept (left) and opposition class short-stay mission (right) (B. Mattfeld and others, 'Trades Between Opposition and Conjunction Class Trajectories for Early Human Missions to Mars' [2014] AIAA SPACE 2014 Conference & Exposition 1. http://www.scopus.com/inward/record.url?eid=2-s2.0-84905996640&partnerID=40&md5=20faabaad7f3b3b0aa7c5c72dee133ad)

over the opposition class mission because conjunction class mission requires overall lower propulsion energy transfer to and from Mars.[27,28]

11.4 Waste Disposal Options for Future Mars Crewed Missions

The requirements for "proper" waste management at any given destination depend on several factors such as number of crew, location, mission duration, available power, re-supply or recycling options, etc. Potential waste management options for crewed space missions should be able to identify and compare feasible waste management options in order to select a suitable waste management system while meeting certain mission-specific criteria.[29]

It is expected that longer duration missions farther away from the Earth into deep space would potentially generate more waste-streams as a result of the distance of the target destination in space and the number of the crew. The Logistics Reduction and Repurposing model predicted a total mass of 8055 kg and a volume of 28.7 m^3 for crew related logistics and the resultant human waste from the consumables amounting

[27] Bret G. Drake, Stephen J Hoffman and Kevin D. Watts, *Human Exploration of Mars Design Reference Architecture 5.0 - Addendum 2* (2009).

[28] David S. F. Portree, 'Humans to Mars: Fifty Years of Mission Planning, 1950–2000' (2001) 21 Monographs in Aerospace History.

[29] S. Doll, B. Cothran and J. Mcghee, 'Selection Criteria for Waste Management Processes in Manned Space Missions.' (1991) 9 Waste Management & Research 345.

to 3840 kg for a trip of a crew of six on a one-year trip to Mars,[30] this is 47.7% of the original logistics mass.[31] Anih et al. (2019) also estimated that without repurposing or recycling of the LRR waste model in a Mars Design Reference Architecture 5.0 (DRA 5.0) mission[32] lasting for a duration ranging from 916 days or more with a crew of six, the waste generated would be close to about 10,000 kg for the period of the mission.[33] Adequate processing of waste during crewed space missions to Mars is therefore very critical to survival of the crew and mission success.[34]

Management of waste-streams during long-duration crewed space missions away from Earth presents various challenges including small habitable volumes, weightlessness, stringent power supply, and high health hazard risk.[35] It is envisaged that methods adapted in previous and current crewed flights as well as other matured space waste management options could be integrated into future crewed space missions. Other waste management options include.

11.4.1 Stowage

Previous and current crewed space missions utilized stowage within the spacecraft as a temporary waste management strategy before final disposal of waste-streams accumulated within the spacecraft.[36,37] Various crewed spacecraft have designated areas where waste-streams can be stored after collection. ISS non-human waste-streams are collected in small containers, stored temporarily in larger Nomex bags, which are removed from the habitable volume[38] and later disposed of using cargo vehicles in the Earth's atmosphere. Longer duration missions away from Earth would

30 James L. Broyan, Michael K Ewert and Patrick W Fink, 'Logistics Reduction Technologies for Exploration Missions', *AIAA Space Conference* (American Institute of Aeronautics and Astronautics (AIAA) 2014).

31 ibid.

32 Drake, Hoffman and Watts (n 32).

33 Samuel Anih and others, 'Waste for Energy and Volume Recovery (WEVR) Using Inductively Heated Plasma Generator', *70th International Astronautical Congress* (2019).

34 Samuel Anih and others, 'Inductively Heated Plasma Generator-Based Waste Decomposition for Long Duration Crewed Space Missions' (2020) ICES-2020- International Conference on Environmental Systems 1.

35 Martin L. Agrella and Malcolm C. Smith, 'Collection, Compaction and Storage of Solid Waste for Space Missions' (1991) 9 Waste Management & Research 365.

36 Michael P. Alazraki and others, 'Solid Waste Management Requirements Definition for Advanced Life Support Missions -Results' [2002] Society of Automotive Engineers.

37 A. E. Drysdale and S. Maxwell, 'Waste System Implications for Mars Missions' (2003) 31 Advances in Space Research 1791.

38 Michael Ewert and others, 'Comparing Trash Disposal to Use as Radiation Shielding for a Mars Transit Vehicle'. https://ttu-ir.tdl.org/ttu-ir/handle/2346/72985.

require effective ways of compacting, stabilizing and storing waste-streams to maximize available space while ensuring crew safety during prolonged stowage within the spacecraft to mitigate off-gassing as well as exposure to hazardous bacteria.

11.4.2 Recycling

Supply and replacement of mission-critical resources, including consumables, becomes more challenging as crewed missions move further away from the Earth and deeper into the solar system. Waste items gradually accumulate within the spacecraft and could pose challenges to the crew if not properly processed and removed from habitable volume. Recycling and reusing will reduce the initial launch mass as well as the trash burden on crewed missions.[39] Recycling is one method of potentially transforming waste into a source of useful supplies for the crew.[40] Various additive manufacturing technology demonstrations on the ISS, such as the one by *Made in Space*, which printed items on the station,[41] and another by *Tethers Unlimited*, an additive manufacturing company which proposed and is testing 3D printing with plastic waste on the ISS.[42] It is envisaged that additive manufacturing will reduce waste load in habitats by providing much needed supplies during transit times and on planetary surfaces from waste materials processed into feedstock.

11.4.3 Bone Yard

Some waste items such as discarded equipment may not be storable within the space habitat but might contain useful spare parts that can be used for the repair of other items. The "bone yard" waste management option is envisaged as an external storage area outside the habitat at locations such as the lunar surface.[43] Since there is a

[39] Michael K. Ewert and others, 'Mission Benefits Analysis of Logistics Reduction Technologies' [2012] 43rd International Conference on Environmental Systems (AIAA-2013-3383) 1.

[40] Leejay Lockhart, 'Recycling in Space: Waste Handling in a Microgravity Environment Challenge' (2018). https://www.nasa.gov/feature/recycling-in-space-waste-handling-in-a-microgravity-environment-challenge. Accessed 11 August 2019.

[41] J. J. Dunn and others, '3D Printing on the International Space Station: Reducing Earth Dependency and Opening New Space Based Markets' (2013) 11 Proceedings of the International Astronautical Congress, IAC 8976. http://www.scopus.com/inward/record.url?eid=2-s2.0-84904648582&partnerID=40&md5=c2bfa3a53e477fff01ea4aa0c97622b8.

[42] Jennifer Harbaugh, 'Refabricator to Recycle, Reuse Plastic Installed on Space Station' (*Humans in Space*, 2019). https://www.nasa.gov/centers/marshall/news/news/refabricator-to-recycle-reuse-plastic-installed-on-station. Accessed 5 January 2020.

[43] NASA, 'Human Integration Design Handbook' (n 10).

possibility of having habitats within tunnels, caves and craters,[44] these geological features could also be likely candidates for bone yard storage. Useful items from damaged or failed equipment could be retrieved and used for repair or replacement of part(s) of an equipment. This option allows for temporary replacement of failed parts pending the arrival of new supplies.

11.4.4 Deep Space Disposal

A spacecraft further away from the Earth could employ deep space disposal which requires ejecting waste from spacecraft into space using designated airlocks. The practicality of trash disposal through airlock has been investigated.

There has been concern about water and other volatiles boiling off of the trash when exposed to vacuum causing movement or the rupture of the trash and inadvertently contaminating inside and outside of the spacecraft due to boil off from water and other volatiles from trash in the airlock thereby affecting the airlock and hindering the ejection of the trash; other concerns include possible contact with the same or other spacecraft and violation of international space treaties through littering in space.[45] Linne et al. (2014) studied crewed missions deep space disposal during missions to Earth-Moon libration points and Mars with attendant ejection challenges, such as the unpredictable behaviour of waste with liquid when exposed to sudden vacuum, the unforeseeable trajectory of the ejected mass as in the case of the libration points near the Moon.[46] Another drawback of deep space disposal includes loss of a finite amount of gas from each airlock opening cycle during waste ejection.

The procedure and technologies implemented from one or more of the above options either as direct or complementary method of waste disposal during future crewed missions to Mars would determine their adherence to planetary protection goals.

11.5 Waste-Streams Impact on Planetary Protection

Unless it is planned for during mission design and provided for in the mission architecture, the trend of leaving waste-stream ‘human signatures’ behind will likely continue during crewed missions to Mars with biological elements and other forms

[44] Jacques Blamont, ‘A Roadmap to Cave Dwelling on the Moon and Mars’ (2014) 54 Advances in Space Research 2140. http://dx.doi.org/10.1016/j.asr.2014.08.019.

[45] Ewert and others (n 43).

[46] Diane L. Linne and others, ‘Waste Management Options for Long-Duration Space Missions : When to Reject, Reuse, or Recycle’ 1. https://ntrs.nasa.gov/archive/nasa/casi.ntrs.nasa.gov/20140010284.pdf.

Fig. 11.4 First picture taken by Neil Armstrong of Apollo 11 on the Moon's surface includes a waste jettison bag on the lunar surface near the lunar module "Eagle" (Eric M. Jones, Ken Glover and Ulrich Lotzmann, 'Apollo Lunar Surface Manual' (2007). https://www.hq.nasa.gov/alsj/alsj-JettBag.html. Accessed 13 July 2019)

of waste, just as the astronauts did during the Apollo crewed missions to the Moon when they dumped their jettison bags filled with waste on the lunar surface (Fig. 11.4).

On April 20, 1967, an unmanned lunar lander, Surveyor 3 touched down near Oceanus Procellarum (Ocean of Storms) on the surface of the moon with television camera among other instruments on board.[47] After two and half years, on November 20, 1969, two Apollo 12 astronauts, Pete Conrad and Alan L. Bean inspected the spacecraft and recovered the TV camera along with other selected components. Upon return to Earth after quarantine, NASA scientists examined the camera and discovered that terrestrial micro-organism (*Streptococcus mitis*) found on the camera were still alive even after exposure to harsh lunar environment.[48] It is therefore imperative to implement proper containment of waste-streams and decontamination of any surface-bound equipment and material before setting foot on Mars.

[47]NASA, 'Analysis of Surveyor 3 Material and Photographs Returned By Apollo 12'. https://www.lpi.usra.edu/lunar/documents/NTRS/collection2/NASA_SP_284.pdf.

[48]F. J. Mitchell and W. S. Ellis, 'Surveyor III: Bacterium Isolated from Lunar-Retieved TV Camera', *2nd Lunar Science Conference* (Lunar Science Institute 1971).

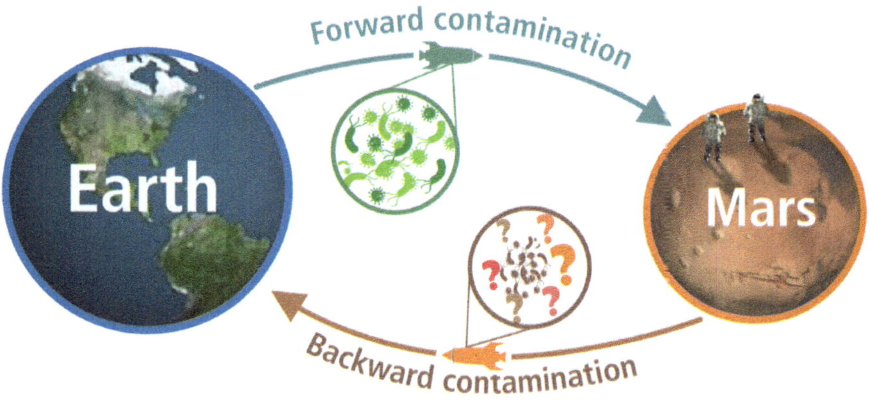

Fig. 11.5 Forward and backward contamination during crewed missions to and from Mars

To prevent both forward and backward contamination (Fig. 11.5), further definition of potential waste streams and contamination pathways will have to be developed including what constitutes "contamination".[49] Also, crewed space missions to Mars should develop and implement procedures for waste and bio-containment to prevent dispersal of potential microbial contamination while on Mars surface and on return to Earth.

One of the drivers for long-duration crewed missions to various other solar system destinations is the search for life. Although a compelling argument can be made for including human astronauts in such missions, clearly, they will be compromised if wastes are not properly managed. For missions, to the Moon and Near Earth Asteroids (NEAs), the threat of contamination is a priority even though these bodies are not expected to sustain life but for missions to potentially life-harbouring destinations, such as Mars, steps must be taken to reduce the risk of contamination.[50] With anticipated increase in Martian population after establishing a permanent settlement on the planet,[51,52] the need to have a sustainable colony will be critical to guarantee possible Earth independence over time.[53]

However, human presence on another celestial body calls for stringent measures to prevent forward and back contamination, especially if the search for life is a

---

[49] John A. Hogan and others, 'Influence of Planetary Protection Guidelines on Waste Management Operations' (2005) 35th Inter SAE Technical Paper Series.

[50] NASA, 'Mission Categories - Protecting Life on Other Bodies' (2017). https://planetaryprotection.nasa.gov/categories.

[51] Jean Marc Salotti and others, 'Impact of Human Factors on the Growing Rate of a Martian Population' (2011) 2 62nd International Astronautical Congress 2011, IAC 2011 1655.

[52] Igor Levchenko and others, 'Mars Colonization: Beyond Getting There' (2019) 3 Global Challenges 1800062.

[53] Samuel Anih, 'Earth and Extra-Terrestrial Sustainable Development: The Challenges of Post-2030 Earth and Space Regime' in Annette Froehlich (ed), *Post 2030-Agenda and the Role of Space* (Springer 2018).

major driver of the crewed mission, given the prevailing conditions on the target body. Future human explorers to Mars have to avoid leaving trails of waste behind them and adopt proper waste management strategies, considering the longer mission duration with the need to maintain a conducive working and living environment for the astronauts.

Different types of waste-streams are expected during crewed space missions with potential for forward contamination of target destinations which depends on so many factors including waste generation parameters and possible contamination channels.[54] Therefore, containment and material dispersion information is required to prevent planetary contamination in line with the Committee on Space Research's (COSPAR) requirements during such missions.[55]

11.5.1 COSPAR Planetary Protection Guidelines

Planetary Protection is the "practice of protecting solar system bodies (planets, moons, comets, and asteroids) from contamination by Earth life, and protecting Earth from possible life forms that may be returned from other solar system bodies".[56] Planetary protection strategies are required at destinations where search for life is of importance such as Mars. COSPAR guidelines[57] builds on the United Nations Article IX of the Treaty on Principles Governing the Activities of States in the Exploration and Use of Outer Space which states: "States Parties to the Treaty shall pursue studies of outer space, including the Moon and other celestial bodies, and conduct their exploration so as to avoid their harmful contamination and also adverse changes in the environment of Earth resulting from the introduction of extra-terrestrial matter, and where necessary, shall adopt appropriate measures for this purpose".[58] Proper waste management during crewed missions to Mars has a critical role to play in planetary protection efforts,[59] from crew operations to habitation on the planet.

The *Planetary Protection Knowledge Gaps for Human Extraterrestrial Missions Workshop Report* identified several Knowledge Gaps and Potential Future Study

[54] Hogan and others (n 55).

[55] NASA, 'NASA Policy on Planetary Protection Requirements for Human Extraterrestrial Missions' 1. http://nodis3.gsfc.nasa.gov/OPD_docs/NPI_8020_7_.doc.

[56] NASA, 'Planetary Protection' (*Office of Safety and Mission Assurance*, 2020). https://sma.nasa.gov/sma-disciplines/planetary-protection. Accessed 15 September 2020.

[57] COSPAR maintains and promulgates this planetary protection policy for the reference of space-faring nations, both as an international standard on procedures to avoid organic-constituent and biological contamination in space exploration, and to provide accepted guidelines in this area to guide compliance with the wording of this UN Space Treaty and other relevant international agreements COSPAR, 'COSPAR Planetary Protection Policy', *COSPAR/IAU Workshop on Planetary Protection* (2011).

[58] United Nations, 'Treaty on Principles Governing the Activities of States in the Exploration and Use of Outer Space, Including the Moon and Other Celestial Bodies', *United Nations Treaty* (1967).

[59] Hogan and others (n 55).

Areas which have to be understood properly to achieve the planetary protection goals.[60] Knowledge Gaps and Potential Future Study Areas describe the challenges for the space exploration community to determine how to prevent contamination during the exploration scenario envisioned within the Journey to Mars framework. This is as a result of insufficient knowledge to determine whether an exploration path as described by the NASA's *Journey to Mars, Pioneering Next Steps in Space Exploration* concept[61] could be performed by human explorers, with the contamination risk of Mars and Earth reduced to acceptable levels.[62]

Three study groups were involved in the study used in the report, with each of the sub-groups indicating how their knowledge gaps is related to current COSPAR implementation guidelines. Altogether, the three study groups were able to identify 25 specific knowledge gaps across the three pre-identified areas of importance.[63] Three main areas of the knowledge gaps identified were used for structuring, analysis and reporting.

The three Knowledge Gaps and Potential Future Study Areas addressed by Workshop Study Groups are *Microbial & Human Health Monitoring*; *Technology & Operations for Contamination Control* and *Natural Transport of Contamination on Mars* summarized which address contamination with respect to crewed missions to Mars. Study Groups 1, 2 and 3 identified nine, eight and eight Knowledge Gaps respectively and discussed them under the following:

> Study Group 1's ***Microbial & Human Health Monitoring*** *identified nine specific knowledge gaps related to microbial and human health monitoring that need to be addressed in order to make progress towards developing quantitative verifiable NASA Procedural Requirement (NPRs).*
>
> Study Group 2's ***Technology & Operations for Contamination Control*** *reviewed technologies required for cleaning, sterilization and prevention of recontamination; mitigation of spacecraft and system effluents; contamination control associated with surface mobility systems and spacesuits; contamination avoidance in Special Regions and* in situ *resource utilization (ISRU) areas; operational strategies to mitigate contamination; and sample containment technologies.*
>
> Study Group 3's **Natural Transport of Contamination on Mars** discussed transport mechanisms on the Mars surface; potential natural sterilization by Martian conditions; and environmental cleanup of inadvertent releases of terrestrial materials.[64]

Planetary protection measures are only effective during crewed space missions to Mars if proper waste-stream management is adhered to. Non-containment of waste-stream within and outside habitats on Mars would expose various locations on the planet to Earth-borne contaminants and instantly nullify attempts to enforce planetary protection. This has implications for Knowledge Gaps 2.7 which specifically

---

[60] Margaret Race and others, 'Planetary Protection Knowledge Gaps for Human Extraterrestrial Missions' (2015). https://ntrs.nasa.gov/archive/nasa/casi.ntrs.nasa.gov/20160012793.pdf.

[61] NASA, 'NASA's Journey to Mars, Pioneering Next Steps in Space Exploration' (n 20).

[62] John D Rummel and others, 'Advances in Planetary Protection Ahead of the "Journey to Mars"' [2016] 46th International Conference on Environmental Systems, Vienna, Austria.

[63] Race and others (n 66).

[64] ibid.

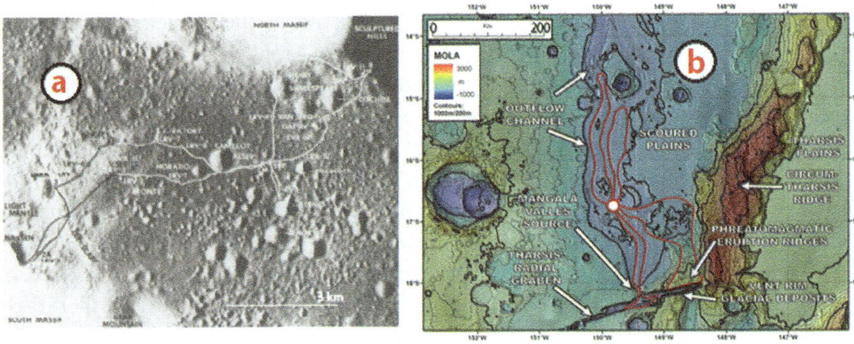

Fig. 11.6 The maximum distance traversed on the Moon by Apollo 17 astronauts was a total of 35.9 km (**a**) (Eric M. Jones and Ken Glover, 'Apollo 11 Image Library' (2017). https://history.nasa.gov/alsj/a11/images11.html) while that for projected traverses for future crewed missions in and around Mangala Valles on Mars (**b**) is far greater (Drake, Hoffman and Watts (n 32)

identified some gaps by asking "*What is "acceptable containment" (type; location; duration) of wastes intentionally left behind? Similarly, what are acceptable constraints and procedures on vented materials*?" This clearly bolsters the need to investigate possible role of waste-streams in contaminating Mars during human exploration.

11.5.2 Planetary Protection and Human Exploration

The COSPAR planetary protection Principles and Guidelines for Human Missions to Mars state that "The greater capability of human explorers can contribute to the astrobiological exploration of Mars only if human-associated contamination is controlled and understood".[65] However, it is anticipated that explorers on Martian surface will have to traverse large expanse of the landing region including other designated sections of the planet and would only be limited by their mobility and life support capability.

Figure 11.6 shows the stark difference between a potential traverse area at Mangala Valles on Mars and the maximum distance traversed on the lunar surface by the *Apollo 17* astronauts. This has a tremendous implication on waste-stream management and possible planetary contamination of Mars in the long run. Atmospheric conditions on the surface of Mars could contribute to waste material dispersal given extensive

[65]COSPAR, 'COSPAR Planetary Protection Policy', *COSPAR/IAU Workshop on Planetary Protection* (2011). http://web.archive.org/web/20130306111646/http://science.nasa.gov/media/medialibrary/2012/05/04/COSPAR_Planetary_Protection_Policy_v3-24-11.pdf.

dust storms and dust devils that are commonplace,[66,67,68] and these could potentially serve as transport mechanisms for various materials including waste-streams from astronauts on the surface. These phenomena could complicate field work and subsequently affect the validity of results obtained in experiments seeking to find evidence of extinct or extant life due to probable contamination of samples.

Moreover, the current stringent planetary protection measures to sterilize spacecraft prior to launch to other planets[69] as well as the strategies for future containment[70] would be instantly nullified if future human visitors to those bodies simply neglect taking proper measures to contain waste-streams within and outside their habitats. Moreover, the viability of waste disposal through incineration by re-entry on Mars would need to be thoroughly investigated to avoid the possibility of also introducing contamination through this route.

11.6 Conclusion

Waste management has become an integral part of crewed space missions unlike spaceflight at the inception of human spaceflight when waste management was an afterthought. Proper waste management has implications for safety of the crewmembers and the success of space missions as is currently being implemented in various spacecraft and on space stations including the ISS. The farthest humans have ventured into the solar system was to the Moon during the Apollo missions to the moon where Apollo astronauts left waste-streams on the moon and it is expected that large number of waste-streams will be generated by the crew during a long-duration mission to Mars as a result all the constraints within the spacecraft and habitats. Therefore, there is the need to develop suitable waste management technologies for managing waste-streams generated within the spacecraft or habitat during the transit phase and on the surface when the crew finally get to Mars. This is because the current mode of waste disposal on ISS would no longer be applicable for that mission. Likewise, astronauts would have to adopt strategies to avoid contaminating the red planet with microbes and other organic materials from waste-streams during such missions. Waste-streams could potentially serve as host for microbes to hitchhike and disperse on the planet as Mars presents peculiar pathways for possible dispersal of waste on the planet due to the active ambient condition on the surface and the possibility of the crew dispersing microbes and other organic materials while traversing further during field

[66]D. A. Waller and R. Greeley, 'Active Dust Devils on Mars : A Comparison of Data Returned from Six Spacecraft Landing Sites.' [2011] Icarus 17.

[67]C. Ho, N. Gloshan and A. Kliore, 'Martian Dust Storms and Their Effects on Propagation', *Radio Wave Propagation for Communication on and around Mars* (Jet Propulsion Laboratory, California Institute of Technology 1999).

[68]Matt Balme and Ronald Greeley, 'Dust Devils on Earth and Mars' (2006) 44 Reviews of Geophysics 1.

[69]COSPAR (n 71).

[70]Race and others (n 66).

work on the surface. Therefore, the potential of waste-streams contributing to the contamination process on Mars do exist however, this requires adequate prevention and mitigation of various waste-streams dispersal to ensure that the signs of life astronauts may likely discover on the red planet will not be those they brought with them from the Earth.

Samuel Anih developed a profound interest in space exploration during his high school days and later founded SpaceRovers which acted as a melting pot for fellow space enthusiasts as an undergrad at Obafemi Awolowo University, Nigeria. He received an M.Sc. from the International Space University (ISU), Illkirch-Graffenstaden (Strasbourg), France and later had a graduate fellowship at NASA Ames, Moffet Field, California where he worked with the Lunar Micro Rover (LMR) team and also proposed an adaptation of NASA's distance learning network platform for collaborative robotic education. He has more than ten years working experience as a scientific officer at the African Regional Centre for Space Science and Technology Education in English (ARCSSTE-E), a United Nations affiliated centre. He is about to complete his Ph.D. from SpaceLab at the University of Cape Town, South Africa

Chapter 12
Shared Network Infrastructures on Mars: Implementing Legal Tools for the Establishment and Regulation of a Martian Power Grid

Robert Bente

Abstract As the colonization of Outer Space becomes a more and more realistic scenario and first intentions to establish human settlements on foreign celestial bodies of our solar system exist, many new legal questions arise. These range from general considerations on the colonization of what is considered a *res communis* to questions addressing specific technologies or ventures. This article shall commit to the latter, yet it necessarily touches several general aspects. It makes a first attempt to briefly sketch some of the manifold legal aspects that might arise during the establishment and development of a Martian power grid. Further, the article takes a position on possible future developments in the legal framework for a permanent human presence on Mars by proposing a draft clause for a possible Mars agreement that sets the basis for a legal and regulatory framework for a Martian power grid.

12.1 Introduction

Space colonization is human settlement in space other than Earth.[1] Space colonization is thus linked to the establishment of human settlements. A human settlement

[1] Rong-En Sun, 'Legal Issues of Space Colonization' (2017) *German Journal of Air and Space Law* 476.

[2] The following definition from the OECD Glossary of Statistical Terms shall serve as an example of the many and in detail varying definitions of the term: 'The term human settlements is an integrative concept that comprises: (a) physical components of shelter and infrastructure; and (b) services to which the physical elements provide support, that is to say, community services such as education, health, culture, welfare, recreation and nutrition.' https://stats.oecd.org/glossary/detail.asp?ID=1266. Accessed 11 October 2020.

Disclaimer: The author is a Referendar at the Regional Court of Leipzig. The information and views set out in this article do not necessarily reflect the opinion of any institution or company the author was, is, or will be affiliated with.

R. Bente (✉)
Regional Court of Leipzig, Leipzig, Germany
e-mail: lawofouterspace@gmail.com

A. Froehlich (ed.), *Assessing a Mars Agreement Including Human Settlements*, Studies in Space Policy 30, https://doi.org/10.1007/978-3-030-65013-1_12

again is any place where people live together in a (more or less) organized community.[2] According to this definition, space colonization as a real act has not begun yet. Literature is thus correctly talking of space colonization as an upcoming topic. However, from an idealistic perspective, it might already have begun with the ISS, humankind's first landing on the Moon, or even the first ratification of the OST,[3] since even the—arguable[4]—decision for a *res communis*[5] may involve an abstract legal act of claiming sovereignty.[6]

After the *Mars One* project, which had officially announced the objective of establishing a permanent human settlement on Mars by 2032,[7] having failed,[8] few approaches to establish human settlements on other celestial bodies than the Earth remain—primarily concentrating on the Moon. One of them is the self-announced space nation *Asgardia*. It is well known for already having launched a satellite into an Earth orbit, claiming the satellite and the space included within it to be a state.[9] But according to its Constitution, *Asgardia* also aims for the expansion of its territory to the Moon and other celestial bodies.[10] Whilst the plans of *Asgardia* are of a very idealistic character, the concept of the *Moon Village*, having been communicated by ESA as a possible future in the exploration of the Moon,[11] forms a more realistic scenario. This idea of a *Moon Village* unveils what might also become an initial scenario for a permanent human presence on Mars.

---

[3]Or at least its entry into force par Art. XIV para. 3 OST (five ratifications) on 10 October 1967.

[4]See Manfred Lachs, *The Law of Outer Space* (Sijthoff Leiden 1972) 48, who doubted outer space and celestial bodies being a '*res*' in this early work.

[5]Ram Jakhu and Steven Freeland, *Cologne Commentary on Space Law*, vol 1 (Stephan Hobe, Bernhard Schmidt-Tedd and Kai-Uwe Schrogl eds, Berliner Wissenschaftsverlag 2017) 225.

[6]Art. II OST only prohibits 'national' appropriation. Declaring space a *res communis* may thus include the assertion that it belongs to mankind; noticing that Art. I OST solely addresses the 'exploration and use' of outer space when speaking of a 'province of all mankind' and does not fully resolve the relationship of mankind to outer space – other than Art. 11 para. 1 MOON ('common heritage of mankind') which directly addresses celestial bodies but still leaves room for interpretation.

[7]After having postponed the date several times, the latest roadmap on the Mars One website schedules a first crew landing for 2032, 'Roadmap'. http://www.mars-one.com/mission/roadmap. Accessed 11 October 2020.

[8]Jeff Foust, 'Mars One company goes bankrupt' (*SpaceNews*, 11 February 2019). https://spacenews.com/mars-one-company-goes-bankrupt/. Accessed 11 October 2020.

[9]*Asgardia-1*, which was launched on 12 November 2017 by Orbital ATK.

[10]See Art. 5 paras. 1, 2 and 6 of 'The Constitution of the Space Nation of Asgardia'. https://asgardia.space/constitution/#c2-a5. Accessed 11 October 2020. Asgardia has also announced the objective of establishing a permanent residence on the Moon until 2043, 'Historic Announcement to All Asgardians: Citizenship Fee'. https://asgardia.space/storage/page/publication/attach/73/8e/738eb39de305e3280d2157d29f62d101e6a4d2ce441d63284af6411deaa5568e.pdf. Accessed 11 October 2020.

[11]Jan Wörner, 'Moon Village: A Vision for Global Cooperation and Space 4.0' (*ESA Blog*, 23 November 2016). https://blogs.esa.int/janwoerner/2016/11/23/moon-village/. Accessed 11 October 2020.

12.1.1 The Need for a Power Grid on Mars

The long distance of a Mars settlement to Earth and the conduction of energy-dependent activities, especially such linked to research, will raise the need for a secure and stable energy supply. Also, in situ resource utilization will require a significant amount of energy.[12] At the beginning of a human settlement on Mars, the number of habitats and other energy-dependent infrastructures may be limited. However, with an increasing number of ventures leading to a growth of the settlement, more and more actors will bring in their infrastructure. This will result in an increasing need for a reliable energy supply.

Like on Earth, in such a scenario, a network solution is favourable to allow the settlement to benefit from network effects, which increase with the number of network users.[13] Also, networks can easily be amended: new users may join, different energy sources may be combined. Within a network structure, every user will indirectly profit from the other users. This provides overall security and reliability to the Mars settlement and minimizes the risk of a total breakdown. Despite this, the availability of a reliable energy network might foster the feasibility of future projects on Mars.[14]

For a Mars settlement, similar thoughts as for the *Moon Village* apply: the idea of a *Moon Village* is to combine the capabilities of multiple space faring nations in a beneficial way.[15] On a smaller scale, this idea applies to a power grid on Mars.

12.1.2 Energy Sources

In general, various energy sources are to consider. This e.g. includes solar power,[16] nuclear power sources, and potentially wind turbines[17] or energy won from the

---

[12]Vidvuds Beldavs, David Dunlop, Jim Crisafulli and Bernard Foing, 'The lunar electrical power utility' (*The Space Review*, 9 November 2015). https://www.thespacereview.com/article/2860/1. Accessed 11 October 2020.

[13]Limited by the heterogeneity of individual preferences for different technologies (network diversity), see Günter Knieps, 'Warum und wozu Regulierung im europäischen Mehr-Ebenen-System? – Gründe für bzw. Ziele von Regulierung' in Ludwig Gramlich and Cornelia Manger-Nestler (eds), *Europäisierte Regulierungsstrukturen und -netzwerke* (Nomos 2011) 25, 27.

[14]Beldavs/Dunlop/Crisafulli/Foing (n 13).

[15]Wörner (n 12).

[16]Noticing that solar power might prove less attractive on Mars because of global dust storm, see Mason Black, 'Powering a Colony on Mars' (*Coursework for PH240, Stanford University, Fall 2017*, 5 December 2017). http://large.stanford.edu/courses/2017/ph240/black1/ (conclusion). Accessed 11 October 2020.

[17]C. Holstein-Rathlou, P. E. Thomas and J. Merrison, 'Wind Turbine Power Production under Current Martian Atmosperic Conditions' (*Mars Workshop on Amazonian Climate 2018*, LPI Contrib. No. 2086). https://www.hou.usra.edu/meetings/amazonian2018/pdf/4004.pdf. Accessed 11 October 2020.

resources of Mars.[18] Out of these, nuclear power sources may only be used subsidiary.[19] In sum, it is to expect that energy supply on Mars will not be linked to one single technology but will include different technologies. This calls for an interoperable solution that is technology-open and capable of adapting to future developments.

12.1.3 Power Plants

Hand in hand with prospective sources of energy goes the question on which kind of power plants will be built on Mars. At this point in time, an answer cannot be given, yet it seems probable that solar power plants as stationary power sources will play a role.[20] For a legal assessment, it might potentially become necessary to differentiate on three levels between

(1) stationary and movable,
(2) ground- and orbit-based, as well as
(3) resource-dependent and renewable energies related power plants.

From all these different categories certain specific legal issues may arise. Nevertheless, legal issues relating to certain energy sources or power plants do not necessarily affect the legal framework applying to the network.

12.1.4 Storage Systems

A special role—not only *in praxi* but also in a legal context—could play storage systems. Depending on the influence energy storages will gain, they might prove as critical parts of the energy network, possibly demanding for special regulation.

---

[18]Because of the lack of specific provisions and since there is no agreement on whether 'free use' in the sense of Art. I para. 2 OST includes the right to take and consume non-renewable natural resources of celestial bodies, it remains unclear whether resource utilization is legal, see Stephan Hobe and Philip de Man, 'National Appropriation of Outer Space and Sate Jurisdiction to Regulate the Exploitation, Exploration and Utilization of Space Resources' (2017) *German Journal of Air and Space Law* 460, 464.

[19]According to Principle 3 of the NPS Principles (UNGA Res 47/68), which is non-binding and not yet *opinio iuris*; Gérardine Goh Escolar and Martin Reynders, *Cologne Commentary on Space Law*, vol 3 (Stephan Hobe, Bernhard Schmidt-Tedd and Kai-Uwe Schrogl eds, Carl Heymanns Verlag 2015) 198f.

[20]Despite the abovementioned concerns, Black (n 17).

12.2 Legal Aspects of Shared Infrastructures on Mars

A Martian power grid, as well as any other shared infrastructure on Mars, raises a myriad of legal issues. These *inter alia* include regulatory aspects and the question, which legal documents may serve as examples for a Mars Agreement.

12.2.1 Existing Legal Documents

A first look shall be taken at existing sources of space law.

12.2.1.1 Moon Agreement

Looking at the sources of treaty-based space law,[21] there is already a treaty applicable to space activities on Mars, as the Moon Agreement applies to any celestial body.[22] Unfortunately, the Moon Agreement has reached only a small number of ratifications till today.[23] But also the sole increase in ratifications would not lead to a much bigger impact unless a sudden very significant increase in ratifications or notably the ratification by the most important space-faring nations would appear. Without a certain number of ratifications and the involvement of major space faring countries, the emergence of customary law is unlikely.[24] Notwithstanding its limited impact, the Moon Agreement is a valuable source of legal approaches to space activities on foreign celestial bodies.[25]

---

[21] Based on the following five treaties: the Outer Space Treaty, the Rescue Agreement, the Liability Convention, the Registration Convention, and – with limitations – the Moon Agreement.

[22] Art. I, para. 1 MOON, stating: 'The provisions of this Agreement relating to the Moon shall also apply to other celestial bodies within the solar system, other than the Earth […]'.

[23] 18 ratifications as of 01 January 2020, UNOOSA, 'Status of International Agreements Relating to Activities in Outer Space as at 1 January 2020'. https://www.unoosa.org/documents/pdf/spacelaw/treatystatus/TreatiesStatus-2020E.pdf. Accessed 11 October 2020.

[24] Relevant factors allowing a metamorphosis to customary law are: a) the lapse of time since the entering into force, b) the number and nature of the respective state parties, and c) whether third states have objected to the conventional rule at stake; according to Valentina Vecchio, 'Customary International Law in the Outer Space Treaty' (2017) *German Journal of Air and Space Law* 491, 494f.

[25] Stephan Hobe, 'The Moon Agreement – Let's Use the Chance' (2010) *German Journal of Air and Space Law* 372, 381.

12.2.1.2 Outer Space Treaty

The OST as the 'magna carta of space law'[26] is the main source of space law. It has been widely ratified,[27] and some of its general provisions already are considered customary international law.[28] Consequently, the provisions of the OST play an important role in legal issues relating to space colonization. Concerning network structures, the provisions of Art. I (freedom of exploration and use),[29] III (accordance with international law), VI (international responsibility), and IX (principle of cooperation and mutual assistance) OST shall be highlighted.

12.2.1.3 Intergovernmental Agreement

An example of successful cooperation of multiple space-faring nations and agencies is the ISS.[30] Its legal framework is based on the IGA[31] and several bilateral agreements.[32] Parts of the IGA may serve as orientation points for the drafting of future international treaties concerning a complex international cooperation, as would be the case with the establishment of a human settlement on Mars. Accordingly, the IGA provisions have already been described by academics as possible guidelines for regulating a *Moon Village*.[33] Especially the provisions of Art. 5, 6, 9, and 10 IGA[34] should be considered for a possible Mars agreement and a power grid clause therein. Further, the IGA makes provisions on a telecommunication network, introducing compatibility requirements for additional networks (Art. 13 para. 1 IGA).

Yet, the vision of a human settlement on Mars causes some major differences to the ISS. Other than the ISS, a settlement and its network infrastructures will not necessarily be limited. Rather, they will be open for new players to join, be it initially,

[26] Francis Lyall and Paul B. Larsen, *Space Law A Treatise* (1st edn, Ashgate 2009) 53.

[27] 110 ratifications as of 1 January 2020, including major space-faring nations, such as the United States of America, the Russian Federation, China, India, and Japan; UNOOSA (n 24).

[28] Vecchio (n 25) 496 ff.

[29] Particularly with regard to section 2, guaranteeing freedom from discrimination and free access.

[30] Setsuko Aoki, 'Analysis of the Legal Instruments Operating the ISS as the Most Complex Space Program Ever Undertaken: From Historical Perspective' in Rafael Moro-Aguilar, P. J. Blount and Tanja Masson-Zwaan (eds), *Proceedings of the International Institute of Space Law 2014* (Eleven International Publishing 2015) 309, 310.

[31] Agreement Among the Government of Canada, Governments of Member States of the European Space Agency, the Government of Japan, the Government of the Russian Federation, and the Government of the United States of America Concerning Cooperation on the Civil International Space Station, TIAS 12927.

[32] Including agreements between governments, agencies, and international intergovernmental organizations, e.g. MOUs between NASA and the respective partners space agencies; Aoki (n 31).

[33] Stephan Hobe and Rada Popova, 'The Moon Village Concept: A Legal Ramification' in P. J. Blount, Tanja Masson-Zwaan, Rafael Moro-Aguilar and Kai-Uwe Schrogl (eds), *Proceedings of the International Institute of Space Law 2018* (Eleven International Publishing 2019) 735, 740.

[34] On registration (Art. 5), ownership (Art. 6), utilization (Art. 9), and operation (Art. 10).

consecutively, or after the respective network has been established. Furthermore, a settlement on Mars will be ground-based and thus raise additional questions in terms of the non-appropriation principle.[35] Much more than the IGA a Mars Agreement requires an open character and must be driven by a (very) long-term vision since a human settlement on Mars might, of course, turn out as a nine-day wonder but might also lead to the formation of a permanently growing human presence, possibly involving many more and currently unknown actors.

12.2.1.4 Other Treaty Law and 'Soft Law'

For specific questions, the LIAB[36] and ARRA[37] may be considered. Besides the five main treaties, many other legal documents with the character of so-called 'soft law'[38] exist that may marginally impact network infrastructures on Mars or the drafting of their legal framework. In this context, the Debris Mitigation Guidelines[39] or the Draft Code of Conduct for Outer Space Activities[40] shall be exemplarily mentioned.

Looking at the developments in lunar exploration, the recently published Artemis Accords[41] provide interesting approaches to be added to the discussion on Mars exploration—especially, since they claim applicability towards any celestial body and particularly Mars.[42]

Further parallels might be drawn to the Law of the Sea or the Antarctic Treaty System. Worthy might also prove a look at international gas pipelines and their

---

[35] Resulting from Art. II OST; compare to the thoughts on installations on the Moon by Hobe/Popova (n 34) 741ff.

[36] Convention on International Liability for Damage Caused by Space Objects, 961 UNTS 187.

[37] Agreement on the Rescue of Astronauts, the Return of Astronauts and the Return of Objects Launched into Outer Space, 672 UNTS 119.

[38] A term that is often criticized, yet widely used, see e.g. Stephan Hobe, *Space Law* (Nomos 2019) 47: 'not very precise'.

[39] Space Debris Mitigation Guidelines of the Committee on the Peaceful Uses of Outer Space, UNGA Res 62/217.

[40] European External Action Service, 'DRAFT International Code of Conduct for Outer Space Activities'. http://www.eeas.europa.eu/archives/docs/non-proliferation-and-disarmament/pdf/space_code_conduct_draft_vers_31-march-2014_en.pdf. Accessed 11 October 2020.

[41] National Aeronautics and Space Administration, 'Principles for Cooperation in the Civil Exploration and Use of the Moon, Mars, Comets, and Asteroids for Peaceful Purposes'. https://www.nasa.gov/specials/artemis-accords/img/Artemis-Accords-signed-13Oct2020.pdf. Accessed 13 October 2020; so far signed by eight countries.

[42] Section 1, subsection 2 of the Artemis Accords, ibid.

regulation[43] or the drawing of parallels to an international smart grid as envisaged by the EU.[44]

As a result, several documents exist that provide relevant material for the drafting of an agreement on Mars activities including human settlement. Though certain of their aspects may be transferred, none of them delivers a comprehensive solution for the implementation of an open network system on a foreign celestial body, and in particular a Martian power grid.

12.2.2 Regulatory Aspects

When considering a Martian power grid, the question arises whether and to what extent early regulation will be necessary. The matter unveils as especially complex since the system of international relations is of network character itself,[45] which requires a multi-level coordination for networks on Mars.

12.2.2.1 General Regulatory Aspects

When it comes to regulation, four main questions must be asked:

(a) What shall be regulated (regulatory scope),
(b) why shall it be regulated (regulatory objectives),
(c) how shall regulation take place (regulatory instruments), and
(d) who shall act as the regulatory authority?[46]

And there are three central objectives of regulation: compatibility and interoperability, provision of universal services, and ensuring a functioning competition.[47]

---

[43]This would particularly include the Directive 2009/73/EC of the European Parliament and of the Council of 13 July 2009 concerning common rules for the internal market in natural gas and repealing Directive 2003/55/EC [2009] OJ L211/94 and the Directive (EU) 2019/692 of the European Parliament and of the Council of 17 April 2019 amending Directive 2009/73/EC concerning common rules for the internal market in natural gas [2019] OJ L117/1; the latter extending the scope of the directive to gas transmission lines to and from third countries.

[44]European Task Force for the Implementation of Smart Grids into the European Internal Market, 'Mission and Work Programme'. https://ec.europa.eu/energy/sites/ener/files/documents/mission_and_workprogramme.pdf. Accessed 11 October 2020.

[45]Sebastian Graf Kielmannsegg, 'Netzwerke im Völkerrecht?' in Sigrid Boysen, Tobias Herbst, Kai von Lewinski and Sabrina Schönrock (eds), *Netzwerke* (47. Assistententagung Öffentliches Recht, Nomos 2007) 83, 98f.

[46]According to Josef Ruthig, 'Europäisierte Regulierungsstrukturen und -netzwerke als Basis einer künftigen Infrastrukturvorsorge' in Ludwig Gramlich and Cornelia Manger-Nestler (eds), *Europäisierte Regulierungsstrukturen und -netzwerke* (Nomos 2011) 11; comparable also Knieps (n 14) 25, who brackets regulatory scope and regulatory instruments as one of thus three basic questions.

[47]Ibid. 26.

Some of these central aspects shall be addressed in the following.

12.2.2.2 Selected Regulatory Aspects of a Martian Power Grid

An early legal approach to the idea of a permanent human presence on Mars provides the opportunity of planning and initially implementing a sustainable and stable network for providing, transferring, storing, using, and generating energy from multiple sources and for various uses.

Regulatory Competency

The lack of national jurisdiction could already prohibit any regulatory approaches on foreign celestial bodies.[48] However, it does not: It may indeed obstruct the regulation of e.g. the allocation of Martian territory.[49] But as far as regulation is concerning a network structure, it only extends to the network itself and the objects forming the physical network structure as well as to the (international) relations between the actors taking part in the network or demanding access towards it.

Access

A major issue with a power grid on Mars might address network access. Since their duplication often is impossible or economically unreasonable, many networks are so called natural monopolies.[50] From the point of achievement of a certain complexity, this will apply to a Martian power grid as well. Therefore, non-discriminatory access to the network for suppliers and users might prove a cornerstone of early network regulation on Mars. Denial of access might de facto make the exploration of Mars impossible, although the free exploration of outer space, including free access to celestial bodies, is guaranteed by Art. I OST and (the largely non-binding) Art. 11 para. 4 MOON. This must be considered especially if for sustainability considerations an obligation to participate in existing network structures should be implemented. Non-discriminatory access is also vital to avoid national monopolisation as prohibited by Art. I paras. 1 and 2 OST[51] and to secure the interest of currently non-space-faring nations, particularly the developing countries, in an equitable involvement. From a network without regulatory measures providing (later) non-discriminatory access, the

[48] According to the principle *nemo plus iuris transferre potest quam ipse habet*, it is irrelevant for this question whether states would regulate themselves or entitle an international agency.

[49] Art. II OST prohibits regarding any part of outer space as part of state territory; Jakhu/Freeland (n 6) 243.

[50] Exemplarily Florian Meister, *Etablierung von Netzwerken in der Energiewirtschaft* (Karl-Friedrich Ackermann and Dieter Wagner eds, Deutscher Universitäts-Verlag 2007) 1.

[51] Stephan Hobe, *Cologne Commentary on Space Law*, vol 1 (Stephan Hobe, Bernhard Schmidt-Tedd and Kai-Uwe Schrogl eds, Berliner Wissenschaftsverlag 2017) 211.

realisation of factual appropriation by excluding others from the necessary energy supply to conduct economically reasonable and competitive activities on Mars might result.

Unbundling

Taking Europe for an example, the energy sector shows one special characteristic: the relevant directives of the EU[52] demand for an unbundling of so-called 'vertically integrated undertakings'.[53] 'Unbundling' in this context means the separation at up to five levels: informational, accounting, operational, legal, and ownership unbundling.[54] European energy regulation thereby focuses on network access and operation of the network.[55] Looking at this example, the question can be raised whether such unbundling should be foreseen for energy-related activities on Mars. The answer is no. The future development of a Martian power grid is open. To allow vertically integrated undertakings can in some cases lead to a better regulatory effect.[56] Any overregulation should be avoided, and unbundling should only be implemented in case a severe demand crystallizes.

Network Operator

The establishment of a network raises one further highly important question: Who shall be the network operator?[57] A new institution, all parties to the Mars Agreement in common, one party that is assigned by the other parties represent a few of the options. As it cannot be foreseen if and how a Martian power grid or other network structures on Mars will develop, the Mars Agreement by itself does not necessarily need to name a concrete network operator or establish a new Agency. To avoid future conflicts, the Mars Agreement still should not ignore this question and oblige the parties to the treaty to name or establish a network operator by the time the network

---

[52]Directive 2009/72/EC of the European Parliament and of the Council of 13 July 2009 concerning common rules for the internal market in electricity and repealing Directive 2003/54/EC [2009] OJ L211/55 for electricity and Directive 2009/73/EC (n 44) for the gas sector.

[53]Par Art. 2 para. 21 Directive 2009/72/EC (ibid.), a vertically integrated undertaking is 'an electricity undertaking or a group of electricity undertakings where the same person or the same persons are entitled, directly or indirectly, to exercise control, and where the undertaking or group of undertakings perform at least one of the functions of transmission or distribution, and at least one of the functions of generation or supply of electricity'.

[54]Kai Uwe Pritzsche and Vivian Vechta, *Energierecht* (1st edn, C. H. Beck 2017) 193.

[55]Stephan Gerstner, 'Der Postsektor als Referenzgebiet europäischer Regulierungsbestrebungen' in Ludwig Gramlich and Cornelia Manger-Nestler (eds), *Europäisierte Regulierungsstrukturen und -netzwerke* (Nomos 2011) 213, 218.

[56]Ibid. 222f.

[57]Similarly, for the use of resources, the question is at stake whether a solution with or without the establishment of an international authority would be favourable; Hobe (n 39) 163.

becomes of significant size. Like discussed for resource utilization, an independent international agency seems favourable.[58]

Pricing

Similar thoughts apply to measurement and pricing of energy quantities. Prospectively, a smart meter installation is imaginable. Even the establishment of an energy stock exchange for Mars, whether as a new stock exchange or affiliated to an existing energy exchange market, seems relatively easy to realize. Such complexity should but be avoided at this early point of time. Nonetheless, when a significant network size on Mars develops, the infrastructure should be ready to react to such development.

Public and Private Actors

Because of the financial expenses required and the (so far) mostly research-oriented character of Mars exploration, mainly state parties are expected to get involved in an early Mars settlement. However, the upcoming 'NewSpace' era[59] suggests that in the future a complex pattern of public and private actors will develop. Private actors are likely to get involved in state activities in the form of public-private partnerships. But exclusively private ventures like a hotel on Mars are also imaginable. Any provisions for a power grid on Mars should therefore be flexible enough to allow a market that will include private entities. Free access to and participation in the network must be open to private actors to allow various types of cooperation and co-existence.

12.2.3 Physical Components

As an outlook, a few aspects relating the components that physically form the network shall be addressed:

[58]Regarding the regulation of space resources, see Ramya Sankaran and Nivedita Raju, 'A Framework to address burgeoning Commercial Complexities in Space Mining ' (2017) *German Journal of Air and Space Law* 70, 98, 102f. and Anita Rinner, 'Space Exploitation – Digging in a Legal Vacuum' in Rafael Moro-Aguilar, P. J. Blount and Tanja Masson-Zwaan (eds), *Proceedings of the International Institute of Space Law 2014* (Eleven International Publishing 2015) 221, 229f.

[59]The term 'NewSpace' wants to reflect the growing commercialization of space activities, see Hobe (n 39) 212.

12.2.3.1 Non-appropriation Principle

Building stationary power plants and other fixed infrastructure on the surface of Mars raises the well-known legal questions on the non-appropriation principle[60] and the exploitation of celestial bodies. Since many other papers specifically address the non-appropriation principle,[61] this will not be subject to this article.

12.2.3.2 Registration/Ownership

Regarding registration and ownership of infrastructure elements, a solution comparable to the ISS[62] and Art. 12 para. 1 MOON seems feasible. At least, a binding and unambiguous agreement of the respective parties must exist[63] to ensure that responsibility for single network installations can easily be allocated to a certain entity. Only then liability or other responsibility issues like a dismantling obligation can become enforceable.

12.2.3.3 Compatibility and Interoperability

Securing compatibility plays an important role in networks.[64] To guarantee compatibility and interoperability of the power grid the network users should agree to common technical standards ensuring the compatibility and exchangeability of components and network parts. The importance of compatibility and interoperability has already been seen by Art. 13 para. 1 IGA and Section 5 of the Artemis Accords, the latter explicitly mentioning power systems besides telecommunication systems as another network relevant infrastructure. Nothing else applies for a Martian power grid. Further, the network should be technology-open (capable of including different or new technologies) and network parts should be interoperable to allow the use of various energy sources.

---

[60] See Brendan Cohen, 'Use Versus Appropriation of Outer Space: The Case for Long-Term Occupancy Rights' in Rafael Moro-Aguilar, P. J. Blount and Tanja Masson-Zwaan (eds), *Proceedings of the International Institute of Space Law 2014* (Eleven International Publishing 2015) 35, 39 ff., who is drawing parallels to the GSO and suggesting to link the use of territory on celestial bodies to the duration of the active use of a site within its original purpose, ibid 45; As for the act of factual appropriation, see Annette Froehlich, 'Utilization - Consumption - Appropriation: Asteroid Mining is in the Pipeline' (2017) *German Journal of Air and Space Law* 268, 273f.

[61] E.g. Andrea Capurso, 'The Non-Appropriation Principle: A Roman Interpretation' in P. J. Blount, Tanja Masson-Zwaan, Rafael Moro-Aguilar and Kai-Uwe Schrogl (eds), *Proceedings of the International Institute of Space Law 2018* (Eleven International Publishing 2019) 111.

[62] Art. 6 para. 1 IGA, clarifying that ownership remains with the providing party.

[63] Comparable to Section 7 of the Artemis Accords (n 42).

[64] Knieps (n 14) 26.

12.2.3.4 Maintenance/Deconstruction

Maintenance could generally stay linked to ownership or responsibility par Art. VI OST. The parties can still agree on a different allocation of maintenance duties by individual agreement. For non-functional infrastructure, a deconstruction obligation should be considered. The article will shortly address this aspect under sustainability aspects.

12.2.4 Sustainability Aspects

Sustainability is a big issue with both the space and the energy sector. Naturally, sustainability aspects will also impact a power grid on Mars.

12.2.4.1 General Sustainability Aspects

General sustainability provisions like the avoidance of harmful contamination, which is explicitly addressed in Art. IX OST and general international environmental law[65] will apply to network infrastructures on Mars as well. In sum, these general provisions might call for obligations to participate in existing networks in order to avoid unnecessary duplications of infrastructure. Also, an obligation to efficient use of energy may result. These general aspects are, however, not specific to a Martian power grid. More specific but still affecting any possible infrastructure on Mars are concerns of dismantling non-functional installations:

12.2.4.2 Dismantling Obligation

Regardless of whether a human settlement on Mars will constantly grow or experience an early cancellation, sustainability considerations must address the question on what shall happen with non-functional or abandoned infrastructure. For orbital debris, the Debris Mitigation Guidelines deliver a few approaches [66] Whenever there is a threat of harmful contamination, dismantling of the threatening infrastructure may become an appropriate measure in the sense of Art. IX OST. However, binding space law very much concentrates on the sustainability of space activities for Earth.[67] But when it comes to significant infrastructure, sustainability concerns relating to foreign

---

[65]On the basis of Art. III OST, which requires compliance with international law.

[66]UNGA Res 62/217 (n 40), especially guidelines 6 and 7, of which only regarding LEO a removal is foreseen.

[67]The only treaty provision demanding for comprehensive environmental protection for foreign celestial bodies is Art. 7 para. 1 MOON since Art. IX OST limits such concerns to harmful contamination. Regarding the limited coverage of Art. IX OST, see Gordon Chung, 'Emergence of Environmental Protection Clauses in Outer Space Treaty: A Lesson from the Rio Principles' in *A Fresh*

celestial bodies gain increasing importance. Space has so far experienced a practice of 'fire and forget'.[68] Similar developments must be avoided during the exploration of foreign celestial bodies. Therefore, a dismantling obligation for non-functional and abandoned infrastructure should be implemented. This also serves to consistently ensure free access (Art. I OST) to the surface of Mars. Since the dismantling of certain parts of a network structure might be dependent on the earlier dismantling of other parts, possibly under the control of a different network user, such obligation may cause some conflicts—especially in cases when the dismantling causes significant costs. The same applies to a dismantling of infrastructure that is still in use or otherwise needed by another network user. Besides the clear allocation of responsibility, this conflict potential indicates a need for an effective dispute settlement.

12.2.5 Responsibility and Liability

Different than on Earth, no boarders could provide for differentiation on the question of who is responsible. This non-territoriality[69] of Mars might prove problematic. Thus, the abovementioned clear allocation of infrastructure parts is vital. Something else might solely apply within 'safety zones' as described by the Artemis Accords in Section 11 no. 7. Looking at Art. 16 IGA,[70] the question can be raised whether a cross-waiver of liability for a power grid on Mars should be agreed on. However, other than for the ISS, a cross-waiver is not suitable for a power grid. It might be a suitable instrument for single projects like the ISS where all concrete participants are known but raises too many insecurities for a network that might indefinitely expand and possibly involve users of different legal nature and with various interests. Too broad and unclear are the liability issues that may arise from a Martian power grid. Thus, general provisions on liability should remain valid. The parties can still include a cross-waiver of liability in a bilateral or multilateral agreement applying to a concrete joint venture.

Very likely parts of the infrastructure will not be built on Earth but by in-situ resource utilization, leading to some problems concerning the applicability of the LIAB. Although, no universal definition of the term 'space object' exists,[71] a space object in the sense of Art. I lit. d LIAB is commonly understood as an object launched

---

View on the Outer Space Treaty (Annette Froehlich ed, Springer International Publishing 2018) 1, 6.

[68] Jan Helge May, 'Space Debris Remediation: Some Aspects of International Law Relating to the Removal of Space Junk from Earth Orbit' (2012) *German Journal of Air and Space Law* 251, 252.

[69] Jakhu/Freeland (n 6) 224.

[70] For the ISS, the cross-waiver is restricted to damage that is a result from the participation in the project, Art. 16 paras. 3 lit. a and 2 lit. f IGA.

[71] Bernhard Schmidt-Tedd and Stephan Mick, *Cologne Commentary on Space Law*, vol 1 (Stephan Hobe, Bernhard Schmidt-Tedd and Kai-Uwe Schrogl eds, Berliner Wissenschaftsverlag 2017) 503f.

into outer space.[72] Following this definition, the LIAB only applies to objects that have once been on Earth. The same results from the concept of the launching state[73] since no launching state in the sense of Art. III LIAB and Art. VII, VIII OST will exist in cases of in situ resource utilization. Art. VI OST, on the other hand, remains unaffected as it does not fail to apply when a launch did not happen. Nonetheless, the *trias* of Art. VI, VII, and VIII OST[74] experiences a breach. These concerns are accompanied by the general uncertainties of the LIAB, including the determination of fault under Art. III LIAB.[75] Despite the obvious deficiencies of the general liability regime in terms of space activities on celestial bodies, a separate liability regime for a power grid does not have to be implemented. Depending on how the general liability framework and potential network structures will develop, some special provisions might become required. However, at this early stage, the task of developing and implementing an effective liability regime is not one to be addressed by a power grid clause. It is rather a matter that must be addressed by general liability provisions. In this context, the international community is asked to agree on a comprehensive and effective liability regime for space activities on foreign celestial bodies, as foreseen by Art. 14 para. 2 MOON.

12.2.6 Duty of Assistance

In parallel to the ARRA and Section 6 of the Artemis accords,[76] a duty of assistance in case of an emergency may be discussed. When a break of energy supply results in a threat to the lives and health of humans, this scenario is likely to be covered by general provisions. Regarding a power grid, the question reduces to the aspect of whether emergency assistance should extend to scenarios in which a break of energy supply threatens tangible assets of another party or the conduction of activities by it.

12.3 Drafting a Power Grid Clause for a Mars Agreement

A possible Mars Agreement should include a clause that addresses shared network infrastructures, particularly a Martian power grid.

[72]Bin Cheng, 'International Responsibility and Liability for Launch Activities' (1995) *Annals of Air and Space Law* 297.

[73]Lesley Jane Smith and Armel Kerrest, *Cologne Commentary on Space Law*, vol 2 (Stephan Hobe, Bernhard Schmidt-Tedd and Kai-Uwe Schrogl eds, Carl Heymanns Verlag 2013) 114.

[74]Compare to Schmidt-Tedd/Mick (n 72) 495.

[75]Smith/Kerrest (n 74) 223.

[76]Both limiting emergency assistance to cases of personnel in distress.

12.3.1 General Aspects of a Mars Agreement

One strength of the OST is its number of ratifications as this eases the establishment of customary law.[77] Of course, a treaty finds its strength primarily in its immediate binding of the parties to the treaty but non-parties to the treaty and the possibility of withdrawal must be given attention to. That is why reaching a high number of ratifications is not only important to directly bind the parties to the treaty but also to foster the emergence of customary law,[78] which will not only bind the parties to the treaty but any other nation, too. This binding character is important because of the long-term scope of the regulations of a possible Mars Agreement—it cannot be foreseen which and how many players will get involved in a settlement on Mars.

Provisions in a Mars Agreement extending to human settlements must be driven by (very) long-term thoughts. Although the space sector is a high technology and thus rapidly changing sector, it is also bound to long-lasting missions and developments, possibly taking decades. The question of a human settlement on Mars could be one of 30, 50, 100, or even hundreds of years. As the OST currently shows, the decisions taken in an early agreement can remain of significant importance even decades later. It is therefore important to implement basic rules for the regulation not only of activities expected within the next years but also of activities that might arise much later.[79]

Similar 'problems' as with the Moon agreement[80] in terms of the arising approaches to exploit the Moon and other celestial bodies (particularly asteroid mining) should be avoided by agreeing to binding basic principles. Any agreement regulating activities on Mars or other celestial bodies, and especially a Martian power grid clause, should thus follow one rule: keep it simple yet binding.

12.3.2 General Aspects of a Power Grid Clause for Mars: Scope and Degree of Detail

Part of the agreement should be general rules enabling a future establishment and regulation of a power grid. Other shared network infrastructures may be addressed as well, giving attention to their respective specifics. Despite all advantages of highly concrete and detailed provisions, a power grid clause in a Mars Agreement should avoid too many details. Overregulation must be avoided. Too many unclarities remain

---

[77] Vecchio (n 25) 494f.

[78] Ibid.

[79] Comparably appealing: Thomas Cheney, 'Developing and Adapting Space Law to Govern Long Term and Permanent Human Settlement of Outer Space, the Moon and Other Celestial Bodies' in P. J. Blount, Tanja Masson-Zwaan, Rafael Moro-Aguilar and Kai-Uwe Schrogl (eds), *Proceedings of the International Institute of Space Law 2018* (Eleven International Publishing 2019) 959, 976.

[80] Meaning its limited practical relevance due to a small number of ratifications, notwithstanding the academic value it provides, Hobe (n 26) 372.

and the implementation of a complete regulatory regime would be inappropriate. Rather, the clause should concentrate on formulating the intents of the parties to the treaty and implement legal tools for future network regulation. Besides that, a few basic provisions, which might prove crucial in the future, should become part of the agreement.

12.3.3 Content of a Power Grid Clause for Mars

A power grid clause should contain provisions on the following aspects.

12.3.3.1 Basic Obligations and Non-discriminatory Access

The clause should outline the general intent of the parties to establish shared infrastructures and guarantee non-discriminatory access towards them. However, it should not prohibit independent infrastructure where appropriate.

12.3.3.2 Technical Standards

To guarantee compatibility and interoperability, the formulation of common technical standards for network components will be key. The Parties should thus be obliged directly by the Mars Agreement to agree to such common standards. However, there is no need to include further details on technical standards. Since technology evolves and standards regularly change, there is rather a need for a flexible solution. Technical standards should therefore be regulated by other legal tools that allow an easy adaption to new technologies and findings.

12.3.3.3 Sustainability Aspects

If not sufficiently regulated elsewhere, the clause should consider sustainability aspects. This particularly involves a dismantling obligation for non-functional or abandoned infrastructure.

12.3.3.4 Network Operator

The clause should open the opportunity of establishing an agency as the network operator and provide an interim solution covering the period prior to the existence of a considerable network.

12.3.3.5 Dispute Settlement

If not sufficiently regulated elsewhere, it should also determine how disputes relating to the clause shall be settled.

12.3.4 Draft Clause

A Mars Agreement should contain a clause to set the legal framework for the establishment and future regulation of a Martian power grid:

'Shared Network Infrastructures, Power Grid

(1) On Mars, the Parties to the Treaty shall establish a shared energy network infrastructure.

(2) Access to such network shall be granted on a non-discriminatory basis and in accordance with the principles of Article I OST. The State Parties to the Treaty shall, by the implementation through national legislation, ensure that non-discriminatory access is guaranteed to any private entity that is subject to the jurisdiction of the respective State Party.

(3) Each Party to the Treaty shall take appropriate measures to foster the development in the sense of section (1). The Parties shall not install infrastructure that will not be connected to the network within reasonable time. This does not affect the right of the Parties to install independent infrastructure where a connection to the network is impossible or unreasonable due to important reasons. The connection of an object to the network shall not affect its ownership or a Party's responsibility for it.

(4) The Parties shall agree on common technical standards for technical network components to secure the compatibility and interoperability within the network and shall comply with these standards. Compliance with the common technical standards shall also be required in cases of independent infrastructure unless a later connection to the network is excluded permanently for factual reasons.

(5) If any network component that is, was, or was supposed to become part of the network is or becomes permanently non-functional or is not expected to become functional again within reasonable time, the Party to the Treaty responsible for this component shall dismantle it and restore the original state that existed before its installation. The obligation to restore does not apply if the component concerned will immediately be replaced by a similar, functional component.

(6) From the time a network has been established that involves more than two Parties to the Treaty, the Parties shall establish an independent agency that shall be responsible for operating the network. They shall also establish an independent body of arbitration, competency to decide on disputes arising from this Article. Until the establishment of such body and in case a diplomatic settlement has failed, the Parties to such a dispute shall bring the dispute before the Permanent Court of Arbitration to settle it in a binding manner.

(7) The provisions of this Article shall also apply to any other shared network infrastructures, taking into account their respective peculiarities.'

12.4 Conclusion

A Mars settlement will result in new challenges and a new level of international cooperation.[81] When a permanent human presence on Mars becomes reality, a reliable energy infrastructure will play a key role. Although it remains unclear what a possible future on Mars will look like, which technical standards will be available, and which players will join the game, it is vital to set a general legal basis to clarify the intents an exploration and colonization of Mars shall be driven by. It is an early point of time, yet a general clause should be implemented in a prospective Mars agreement to provide a basic framework for the future establishment and regulation of a Martian power grid. The clause may also extend to other networks. It should implement legal tools for regulation and already contain provisions on network access and sustainability issues. However, it should not implement a highly detailed regulation since the future demand for regulation remains unknown and regulation in detail must be specific to the respective network type and status. A Mars Agreement can insofar have the character of a large-scale preamble. The OST has shown how beneficial an early approach to regulate the unknown can be.

Robert Bente studied law at the University of Leipzig and is currently a Referendar at the Regional Court of Leipzig (Germany). Besides his studies, he has been supporting the Chair of Civil Law, German and International Business Law at the University of Leipzig as well as a law firm with a focus on renewable energies as a student and research assistant. Since obtaining the First State Exam the author has begun to extend his interest towards the space sector, recently completing the 'Verwaltungsstation' at the German Federal Ministry of Transport and Digital Infrastructure where he worked on issues related to the Galileo programme of the European Union.

[81] As already concluded by Sun (n 2) 490.

Chapter 13
A Tale of Two Planets in International Space Law: Limitations to the Freedom of Exploration and Use

Radhey Soundarya Gnanesh

Abstract Since time immemorial, humankind has held a deep fascination for the stars and the skies that lay beyond our planet. It is these boundless realms of promise that propelled the space race that resulted in humankind's first foray into lunar terrain. Now several decades later, the world has already begun its preparation for its next big leap into space. Establishing a multi-planetary existence, previously only far fetched science fiction, has become a real possibility with governments and private space enterprises alike, setting their sights on colonizing Mars by the end of the century. This increased attention towards the last frontier, has given rise to a number of problematic concerns, particularly in the legal and policy sphere. The exploration of Mars, at all stages involves the extraction and use of Martian resources. Such activities undertaken by singular entities in the context of the protections afforded to the international community at large, leads to complex legal issues. If economic exploitation is indeed permitted by the Outer Space Treaty, then how will this impact the interests of other nations, particularly the non-spacefaring nations? This article focuses on these pivotal legal aspects, the entailments of the freedom of 'use' as stated in Article II of the Treaty and the viability of limitations imposed by treaty law upon the exercise of this freedom. While examining the above dimensions, this article also brings forth issues of sufficiency of the present legal framework, and discusses its ability to provide for complex issues arising from future space settlements. It addresses the key question of whether amendments are necessary to the existing structure or if an entirely new treaty should be adopted to provide for the legal niches of space settlements, in particular the colonization of Mars.

13.1 Extraction and Utilisation of Resources in Outer Space

A new space race, this time to extract resources from outer space, seems to be afoot. Countries are trying to keep ahead of each other by sending probes to the moon to

---

R. S. Gnanesh (✉)
Gujarat National Law University, Gandhinagar, Gujarat, India
e-mail: radhey1997@gmail.com

A. Froehlich (ed.), *Assessing a Mars Agreement Including Human Settlements*, Studies in Space Policy 30, https://doi.org/10.1007/978-3-030-65013-1_13

map the Moon's resources, and identify the quantities that it is available in.[1] In the past decade alone, India,[2] China[3] and the United States,[4] among others, have already sent probes to the moon that either have or have had equipment to specifically map its resources.[5] For instance, China at present has a three-phase moon exploration plan that is both thorough and ambitious,[6] with plans to design a fourth phase.[7] It's recently launched Chang'e 5 is equipped with a Lunar Regolith Penetrating Radar,[8] that is aimed at excavating lunar terrain, for the purpose of returning samples.[9]

With regard to Mars as well, the prospect of its colonization has lead to a new wave of research into Martian resources and techniques for its extraction, as they may well prove to be a means of sustenance for life on the Red Planet. So far, meteorites from Mars have been found to have valuable elements like magnesium, aluminium, titanium, iron, chromium and trace elements like nickel, copper, zinc, niobium, molybdenum, lanthanum, europium, tungsten, and gold. Given this, it is quite possible that the above-mentioned minerals are available in minable quantities, as well.[10] In addition to this, elements that are of life sustaining nature, such as hydrogen, oxygen and water are also sought to be extracted. For this purpose, NASA has already developed the Regolith Advanced Surface Systems Operations Robot or 'RASSOR 2.0', which is a mining robot that can dig down to the regolith on the surface of Mars, extract these elements and transfer them to a processing plant where they can then be used for life support systems.[11]

---

[1] Byron C. Brittingham, 'Does the World Really Need New Space Law?' (2010) 12 OR REV INTL L 31.

[2] 'Chandrayaan 2: Expanding the Boundaries of Human Knowledge' (Indian Space Research Organisation). https://www.isro.gov.in/chandrayaan2-home-0. Accessed 19 July 2020.

[3] Leonard David, 'On the Far Side of The Moon, Chinese Lander and Rover hit One-Year Mark' (Space.com, 3 January 2020). https://www.space.com/china-moon-far-side-lander-rover-chang-e-4-milestone.html. Accessed 19 July 2020.

[4] 'What is LCROSS, the Lunar Crater Observation and Sensing Satellite?' (National Aeronautics and Space Administration, 8 March 2019). https://www.nasa.gov/ames/lcross. Accessed 19 July 2020.

[5] Byron C. Brittingham, 'Does the World Really Need New Space Law?' (2010) 12 OR REV INTL L 31.

[6] Ram S. Jakhu, Joseph N. Pelton and Yaw O. Mankata, *Space Mining and its Regulation* (Springer International Publishing 2017) 106.

[7] Xinhua, 'China to Launch Chang'e 5 Lunar Probe in 2020' (Xinhua Net, 26 October 2019). http://www.xinhuanet.com/english/2019-10/26/c_138505684.htm. Accessed 19 July 2020.

[8] 'Future Chinese Lunar Missions' (National Aeronautics and Space Administration). https://nssdc.gsfc.nasa.gov/planetary/lunar/cnsa_moon_future.html. Accessed 19 July 2020.

[9] Xinhua, 'China to Launch Chang'e 5 Lunar Probe in 2020' (Xinhua Net, 26 October 2019). http://www.xinhuanet.com/english/2019-10/26/c_138505684.htm. Accessed 19 July 2020.

[10] Wilmer Giraldo and Jorge I. Tobon, 'Extraterrestrial Minerals and Future Frontiers in Mineral Exploration' (2013) 80 DYNA 83.

[11] Anna Heiney, 'Rassor, Marco Polo Demonstrate Resource Utilisation on Mars' (National Aeronautics and Space Administration, 3 October 2016). https://blogs.nasa.gov/kennedy/2016/10/03/rassor-marco-polo-demonstrate-resource-utilization-on-mars/. Accessed 19 July 2020.

There seems to be no dearth in private initiatives either, with Space X testing prototypes of the Starship which is a 'fully reusable transportation system designed to carry both crew and cargo to the Moon, Mars and beyond.'[12] A company focused on establishing human settlements on Mars, called 'Mars One', was established at the start of the decade. It is presently in the process of selecting four people, who are to undertake a one-way journey to the Red Planet, as early as in 2031.[13]

However, despite these developments, the activity of mining in outer space is still a contested subject in international law and scholarship. This is even more so, given that there exists no specific treaty or regulations that govern only this aspect of space activities. This is unlike the activities of remote sensing in outer space[14] and mining in another global common, the deep seabed,[15] both of which have particular regulations governing them. Therefore, this naturally brings the question of permissibility of mining in outer space under the present framework of international law. For this it becomes necessary to have recourse to the provisions of the Outer Space Treaty[16] [*hereinafter,* "OST"], several of whose provisions are considered by the majority of scholars to represent customary international law.[17]

13.2 Freedom of Exploration and Use of Outer Space

Articles I, II and III along with the preamble of the OST were principles that were codified, for the most part from the Declaration of Legal Principles.[18] There was very little disagreement on the above-mentioned provisions among the members of the Legal Subcommittee, who were entrusted with the preparation of the final treaty on Outer Space. The freedom of 'exploration and use' of outer space, including the Moon and other celestial bodies, is viewed as an essential aspect of international space law and is enshrined in Article I of the OST. This provision is essentially a

---

[12]Starship Update (SpaceX). https://www.spacex.com/vehicles/starship/. Accessed 20 July 2020.

[13]Mars One Mission Roadmap (Mars One). http://www.mars-one.com/mission/roadmap. Accessed 20 July 2020.

[14]UNGA Res 6621 (1986) GAOR 21st Session Supp 16.

[15]United Nations Convention on the Law of the Sea (adopted 10 December 1982, entered into force 16 November 1994) 1833 UNTS 397 (UNCLOS).

[16]Treaty on Principles Governing the Activities of States in the Exploration and Use of Outer Space, Including the Moon and Other Celestial Bodies, (adopted 19 December 1966, entered into force 10 October 1967) 610 UNTS 205 (Outer Space Treaty) art 1.

[17]Thomas Gangale, Marilyn Dudley-Rowley, 'To Build Bifrost: Developing Space Property Rights and Infrastructure' (2005) American Institute of Aeronautics and Astronautics, Working Paper 8. http://www.astrosociology.com/Library/PDF/Submissions/To%20Build%20Bifrost.pdf. Accessed 20 July 2020.

[18]UNGA Res 5656 (1963) GAOR 18th Session Supp 15.

microcosm of the entire treaty in that, it grants freedoms for certain activities, which it then regulates by stating limitations to the same by way of its other provisions.[19]

The limitations to the freedom of 'exploration and use' exist primarily in Article I and II of the OST, and will be discussed in the course of this article.

13.2.1 Freedom of 'Exploration'

At the outset, it must be noted that the 'freedom of exploration' is different from the freedom to carry out 'scientific investigation', also contained in Article I of the OST. The first encompasses the second as the term 'exploration' when understood in its literal sense, refers to the general finding out of something that is yet to be explored, which may or may not include scientific activity. Thus, this freedom coupled with the freedom of access as stated in the same provision, essentially means that all States have a right to engage in the exploration of all parts of outer space, including the Moon and other celestial bodies.

13.2.2 Freedom of 'Use' of Outer Space

With regard to the freedom of 'use' of outer space, including the Moon and other celestial bodies,[20] unlike UNCLOS,[21] the OST does not give a particularly detailed explanation as to the entailments of the term 'use'.[22] It is a well-established practice in the interpretation of treaty law that in the event of any ambiguities in any provisions, the Vienna Convention of the Law of Treaties[23] [*hereinafter*, "Vienna Convention"] is rendered applicable. As per Article 32 of the Vienna Convention, reference may be had to the *travaux préparatoires* of the treaty, in order to address any ambiguities.

During the negotiations leading up to the formation of the final and approved treaty, there was some relatively controversial discussion as to the meaning of the term 'use'

[19] Stephan Hobe, 'Article I' in Stephan Hobe, Bernhard Schmidt-Tedd, and Kai Uwe Schrogl (eds), *Cologne Commentary On Space Law*, vol 1 (Carl Heymanns Verlag 2009) 27.

[20] Treaty on Principles Governing the Activities of States in the Exploration and Use of Outer Space, Including the Moon and Other Celestial Bodies, (adopted 19 December 1966, entered into force 10 October 1967) 610 UNTS 205 (Outer Space Treaty) art 1 para 2.

[21] United Nations Convention on the Law of the Sea (adopted 10 December 1982, entered into force 16 November 1994) 1833 UNTS 397 (UNCLOS).

[22] Irmgard Marboe and Christopher D. Johnson, 'The Basic Points of International Space Laws with regard to Resource Mining' in Stephan Hobe (ed.), Background Paper: Does International Space Law Either Prohibit The Taking Of Resources In Outer Space and On Celestial Bodies And How Is This Relevant For National Actors? What Is The Context And What Are The Contours And Limitations Of This Permission Or Prohibition' (IISL Directorate of Studies 2016) 30. https://iislweb.org/docs/IISL_Space_Mining_Study.pdf. Accessed 6 September 2020.

[23] Vienna Convention on the Law of Treaties, (adopted on 23 May 1969, entered into force 27 January 1980) 1155 UNTS 331 (VCLT) art 32.

as it exists today in the provision. France, as part of this discussion, while referring to the existing uses of outer space for purposes such as meteorological research and telecommunications, also referred to the extraction of minerals as being a potential use of the Moon.[24] It was in this context, that France proposed that the term 'use' means exploitation. Most states agreed with France on this understanding of 'use',[25] thereby lending support to a broad interpretation of the term. This is also evidenced by the usage of the term 'use' subsequent to the phrase 'exploration and exploitation' in the United Nations General Assembly Resolution on the subject.[26] Thus, the term 'use' can comfortably be taken to include with its ambit, 'exploitation' as well.

13.2.2.1 Scope of Permitted 'Use'

With regard to its scope, the freedom of 'use' has generally been accepted as including within its scope, both economic and non-economic use.[27] An example of 'use' of outer space in practice is the well-recognized use of frequencies and orbital positions of satellites, both for commercial and non-commercial purposes by several States. Thus, the use of outer space can very well include the exploitation of extraterrestrial resources with the objective of making economic profit, among other things.[28] This provides a legal basis for the extraction and mining of resources as a permissible activity in international law, under the auspices of the OST itself. This is also supported by the position taken by the majority in the 'S.S. "Lotus"' case according to which, 'restrictions upon the independence of states cannot be presumed.'[29] This dictum has been understood as the permissive principle, which means that unless something is prohibited per se under international law, it cannot be regarded as illegal.[30] When applied to space treaty law, the understanding that specific uses cannot be regarded as impermissible unless explicitly prohibited by the provisions of the OST,[31] acts as an additional support to this interpretation.

Therefore, this freedom can be exercised by governments and private space enterprises acting under the 'authorization and supervision' of their governments. This is as

[24]UNCOPOUS Legal Sub Committee (5th Session) 'Summary Record of Sixty Third Meeting' (1966) U.N. Doc A/AC.105/C.2/SR.63, 8; UNCOPOUS Legal Sub Committee (5th Session) 'Summary Record of Sixty Ninth Meeting' (1966) U.N. Doc A/AC.105/C.2/SR.69, 5.

[25]Stephan Hobe, 'Article I' in Stephan Hobe, Bernhard Schmidt-Tedd, and Kai Uwe Schrogl (eds), *Cologne Commentary On Space Law*, vol 1 (Carl Heymanns Verlag 2009) 30.

[26]UNGA Res 1348 (XIII) (1958) GAOR 13th Session Supp 18, 18A.

[27]Stephan Hobe, 'Article I' in Stephan Hobe, Bernhard Schmidt-Tedd, and Kai Uwe Schrogl (eds), *Cologne Commentary On Space Law*, vol 1 (Carl Heymanns Verlag 2009) 35.

[28]Stephan Hobe, 'Article I' in Stephan Hobe, Bernhard Schmidt-Tedd, and Kai Uwe Schrogl (eds), *Cologne Commentary On Space Law*, vol 1 (Carl Heymanns Verlag 2009) 35.

[29]The Case of the S.S. "Lotus" (France v Turkey) Judgment (1927) PCIJ Series A no 10, 18.

[30]Prosper Weil, '"The Court Cannot Conclude Definitively…": *Non Liquet* Revisited' (1998) 36 COL J TRANSNATL L 109, 112.

[31]Stephan Hobe, 'Article I' in Stephan Hobe, Bernhard Schmidt-Tedd, and Kai Uwe Schrogl (eds), *Cologne Commentary On Space Law*, vol 1 (Carl Heymanns Verlag 2009) 35.

per Article VI of the OST, which requires that the activities of non-governmental entities in outer space require "authorization and continuous supervision by the appropriate State Party to the OST."[32] This, at present may be by way of a licensing regime, among other means. The drafters of what is often considered the mother treaty of international space law agreed on the principle of 'private activity but public responsibility' and in doing so, ensured the preservation of both the US liberalist thought and the sovereign ideals of the former USSR.

This freedom of 'exploration and use', like all freedoms given, carries with it a certain set of limitations, which indicates the caution with which these freedoms are to be exercised. Accordingly, if states must exercise these freedoms, it must be with utmost care while keeping in high regard, the benefits and interests of all nations and the international community as a whole. The limitations to the freedoms detailed above, are the principles of 'common benefit of all nations', 'province of all mankind', 'non-discrimination' and 'non-appropriation', which are contained in Article I and II of the OST. These limitations will be addressed provision wise by this article.

13.3 Limitations Contained in Article I

As per Article I, paragraph 1 of the OST,

> The exploration and use of outer space shall be carried out for the benefit and interest of all countries irrespective of their degree of economic and scientific development and shall be the province of all mankind.[33]

Upon a close inspection of the above paragraph, three limitations to the freedom of use and exploitation can be identified. The first limitation is the common benefit and interest principle, the second is the non-discrimination principle and the third is the 'province of all mankind' principle. The implications of the freedom to exploit resources on non-space faring nations would be evident upon the discussion of these limitations.

13.3.1 Common Benefit and Interest

The central idea behind this concept is that, the benefits of activities of space exploration must not be exclusive to one country, and must rather be in the interest of all

[32] Treaty on Principles Governing the Activities of States in the Exploration and Use of Outer Space, Including the Moon and Other Celestial Bodies, (adopted 19 December 1966, entered into force 10 October 1967) 610 UNTS 205 (Outer Space Treaty) art 6.

[33] Treaty on Principles Governing the Activities of States in the Exploration and Use of Outer Space, Including the Moon and Other Celestial Bodies, (adopted 19 December 1966, entered into force 10 October 1967) 610 UNTS 205 (Outer Space Treaty) art 1 para 2.

states, whether or not that State is part of that activity. It attempts to ensure equitable sharing of these benefits among all states, by providing that even the nations that have incipient space programmes or are non-space faring should benefit from the results of space activities. Interestingly enough, this provision is reflective of the inclinations of some developing countries as well.[34] This concept is one of the guiding cornerstones of the OST, and is stated even in its preamble. However, although the preamble does possess binding force as it indicates the aspirations of the drafters of the treaty, its words still cannot create commitments beyond the actual text of the treaty.[35] Therefore, reliance would be more viable, if placed only on the text of the treaty itself. Despite the clear guiding principles behind this provision, the phrase by itself is quite vague and therefore unclear in its manner of application to actual State activities. In such situations, it is useful to refer to subsequent practice by States, to form a practical understanding of its application.[36]

The approach of States to the above principle has been unfulfilling of its aspirations, as such a provision which implies a need for 'equitable sharing' among States, is viewed as unfairly advantageous to the States with no or developing space programmes. This is enunciated by the presence of the UNGA Declaration on Space Benefits[37] where it is made clear that there exists no general obligation that binds the space faring nations to cooperate in this regard. Therefore, there is no general duty upon nations with space capabilities to grant benefits to nations with no or developing space programmes, unless on mutually acceptable terms. This emphasis on 'mutually acceptable terms' is seen throughout the declaration and implies that, if at all such benefits are sought to be acquired by these nations, there must be corresponding efforts at the bilateral level, whereby an understanding is reached between the space faring and the nations with emerging or no space capabilities for equitable sharing of benefits. This system of bilateral agreements, allows for the freedom of the space faring nation to decide if entry into such an agreement, suits its needs, thereby in essence, allowing a situation that may be entirely contrary to the above stated principle.

Regardless of the general policy implications of the above, it remains quite clear that space faring nations, by undertaking activities of exploration and exploitation in outer space, do not automatically become bound to share the profits with the non-participating nations by virtue of the treaty which perhaps if done, may ensure a direct implementation of the common benefit principle. That is not to say that there is no benefit by their activities to other nations, including non-space faring nations at all, as there is indeed an indirect benefit derived from such activities by the international community at large. The discoveries and advancements in space technologies made

---

[34]Stephan Hobe, 'Article I' in Stephan Hobe, Bernhard Schmidt-Tedd, and Kai Uwe Schrogl (eds), *Cologne Commentary On Space Law*, vol 1 (Carl Heymanns Verlag 2009) 41.

[35]Hans-Dietrich Treviranus, 'Preamble' in Rudolph Bernhardt (ed), *Encyclopedia of Public International Law*, vol 3 (Elsevier 1997) 1097.

[36]Christian Djeffal, 'Commentaries on the Law of Treaties: A Review Essay Reflecting on the Genre of Commentaries' (2013) 24 (Issue 4) EUR J INTL L 1223.

[37]UNGA Res 51/122 (1996) GAOR 51st Session Supp 49.

by space faring nations, would in fact allow for presently non-space faring nations to engage in space activities in the future. This is due to the reasonable deduction that along with development, more viable services will become available at cheaper costs. For instance, the carrying costs of payloads have reduced significantly since the first payloads that were carried into space. Additionally, if a cheaper energy source or more cost effective aerospace engineering methods are found, that would in fact make space accessible for all nations.

Therefore, to sum the applicability of this limitation, all activities in outer space including extraction and mining still can be regarded as being carried out for the benefit and in the interests of all countries, without there existing a resource or profit sharing regime per se. This is in accordance with the general guiding principles of common benefit to and non-discrimination among all nations, 'irrespective of their degree of economic or scientific development'. The activities of exploitation and resource utilization are thus not hindered by this limitation, from a legal and practical standpoint.

13.3.2 Freedom of Access to All States

This freedom in the specific context of resource utilization, gives rise to complex questions with regard to the exclusivity of use of particular resources in a defined territory. The expansive nature of the phrase 'discrimination of any kind', makes it clear that there must not be any differential treatment between states on the basis of technological advancement as well. However clear this may be in theory, the freedom of access provided to all states in the background of the non-discrimination principle, does present some perplexing questions when applied in practice.

For instance, consider a hypothetical situation where a country engages in a resource mining expedition and removes resources from a particular area. Does the question of whether this activity inherently imposes upon another country's ability to acquire resources from the same area, not naturally arise? This almost automatic question is premised on two levels of reasoning. First, if a country 'X' mines in a defined area, plot 'A' at a particular point in time, another country, 'Y', for obvious reasons cannot extract the same target resources from the same area at the same point in time. This could be understood as precluding in a sense, country 'Y' from carrying out activities on the same plot 'A' at the same time. This is due to a reasonable impossibility of multiplicity of activities by two states on the same exact area, plot 'A', supported by the practical understanding that two countries would not want to share the same place to carry out their activities, for at the very least, political and security reasons. Does this restriction not then adversely impact the 'freedom of access' that is so unequivocally granted to all countries and which allows access to *all* parts of the space, *including* the very same plot A? Second, given that many State parties to the

OST do not presently have independent space faring abilities,[38] another contentious area arises. Does the extraction of resources by a particular country from a defined area inherently limit a presently non-space faring nation to acquire those very same resources, at a later point in time, when they become technologically able?

These are pertinent questions that must be answered at the earliest, after extensive multilateral discussions on the same, so as to arrive at a common understanding on the international community's stance on these complex legal considerations prior to embarking on the journey to the next frontier. One thing however remains definitive at this point. In a hypothetical situation, if *all* the resources on a particular body are utilized, then other nations cannot reasonably utilise any resources on that body any longer.[39] This is a clear violation of the non-appropriation principle contained in the OST.[40] The rest of the questions as outlined above, remain to be decided by the international community by way of discussion and cooperation, so that a consensus may be arrived at.

13.3.3 Principle of Non-appropriation

The principle of non-appropriation as contained in Article II of the OST, aims to confer the status of *res communis omnium* to the realm of outer space.[41] However, there are reasons to support that this prohibition existed even prior to the OST coming into force, by virtue of it already being a part of customary international law.[42] The unanimous adoption of the UNGA resolutions[43] that stated this principle in essentially the same terms as the OST, prior to the adoption of the OST, points to the *opinio juris* that outer space was never intended to be subject to appropriation.[44] This accepted prohibition then was specifically included in the text of the OST to guarantee the protection of outer space from national or exclusive colonization by States. The prevention of the exclusive colonization of outer space by a State was a

[38] 'United Nations and Republic of Korea Workshop on Space Law United Nations Treaties on Outer Space: Actions at the National Level' (United Nations Office of Outer Space Affairs). https://www.unoosa.org/pdf/publications/st_space_22E.pdf. Accessed 3 August 2020.

[39] Stephan Hobe, 'Article I' in Stephan Hobe, Bernhard Schmidt-Tedd, and Kai Uwe Schrogl (eds), *Cologne Commentary On Space Law*, vol 1 (Carl Heymanns Verlag 2009) 42.

[40] Stephan Hobe, 'Article I' in Stephan Hobe, Bernhard Schmidt-Tedd, and Kai Uwe Schrogl (eds), *Cologne Commentary On Space Law*, vol 1 (Carl Heymanns Verlag 2009) 42.

[41] Fabio Tronchetti, 'The Non-Appropriation Principle Under Attack: Using Article II of the Outer Space Treaty in Its Defence' (2007) 50 PROC L OUTER SPACE 526, 530.

[42] Steven Freeland, Ram S. Jakhu, 'Article II', Stephan Hobe, Bernhard Schmidt-Tedd, and Kai Uwe Schrogl (eds), *Cologne Commentary On Space Law*, vol 1 (Carl Heymanns Verlag 2009) 53.

[43] UNGA Res 1721 (XVI) (1961) GAOR 16st Session Supp 17; UNGA Res 1962 (XVIII) (1963) GAOR 18th Session Supp 15.

[44] Fabio Tronchetti, 'The Non-Appropriation Principle Under Attack: Using Article II of the Outer Space Treaty in Its Defence' (2007) 50 PROC L OUTER SPACE 526, 530.

primary concern at the time of its preparation, understandably given the Cold War origins of the OST.

It is undisputable that this principle has served as a useful tool in, holding nations back from establishing areas of exclusive use and occupation in outer space and thus, also preventing military advances into outer space.[45] In recent times, given the rapid commercialization of outer space, the scope and relevance of this principle has been subject to both direct and indirect attack. There is discussion in legal scholarship that even calls for its removal,[46] or at the very least, an amendment on the premise that this provision in its original form, no longer serves a nation's interest in the present times. This article after tracing the contours of the principle in the context of the various glaring issues it gives rise to, in this era of commercialization and exploitation of outer space, will put forth its suggestions on the same.

With regard to territorial claims in outer space, it is clear from the inclusion of several phrases including the phrase, 'province of all mankind', that a claim of sovereignty over land is prohibited.[47] Apart from this, it was made clear during the drafting of the treaty that celestial bodies are not to be subject to *any* claim of sovereignty. This understanding is supported by the *travaux préparatoires* of the OST, which includes a letter from Arthur Goldberg, the Permanent Representative of the United States to the Chairman of the Committee on Peaceful Uses of Outer Space[48] as well as the negotiating history of the Treaty, where statements made by the Belgian, Brazilian and the Australian delegates point at this prohibition.[49] Even with regard to the Moon Agreement, a similar understanding was found among the delegates that any commercial or exploitative activities require the consent of the community of States.[50] This also implies naturally that such prohibited claims of sovereignty cannot be validated by the mere enactment of a national legislation either.

Despite the above, there lie the two related questions of appropriation and ownership of outer space territory and resources, the answers to which have become unclear after the practice exhibited by States after the conclusion of the OST. For instance, the moon rocks collected by the crew of the Apollo 11 missions by the United

[45] Kemal Baslar, *The Concept of The Common Heritage of Mankind in International Law* (Martinus Nijhoff 1998) 91.

[46] Kurt A. Baca, 'Property Rights in Outer Space' (1993) 59 J AIR LAW AND COM 1041; Glen H. Reynolds, 'International Space Law: Into the Twenty-First Century' (1992) 25 VAND J TRANSNATL L 225.

[47] Mary V. White, 'The Common Heritage of Mankind: An Assessment' (1982) 14 CASE W RES J INTL L 509, 535.

[48] Letter from Arthur Goldberg, Permanent Representative of the U.S., to the Chairman of the Comm. on the Peaceful Uses of Outer Space (June 16, 1966), http://www.unoosa.org/pdf/limited/c2/AC105_C2_L012E.pdf. Accessed 3rd August, 2020.

[49] UNCOPOUS Legal Sub Committee (5th Session) 'Summary Record of Seventy First Meeting' (1966) U.N. Doc A/AC.105/C.2/SR.71 and Add 1.

[50] Frans G. von der Dunk, 'The Moon Agreement and the Prospect of Commercial Exploitation of Lunar Resources' (2007) 32 ANNALS AIR AND SPACE L 91, 98.

States, as with all Apollo Missions are explicitly stated as being the 'property of the United States'.[51] This was followed by the enactment of a domestic legislation, permitting the mining of asteroid bodies, and declaring ownership of the resources so mined.[52] Other major space faring nations such as Japan,[53] have already followed suit, while China has issued government regulations[54] in this regard and the United Arab Emirates[55] is due to enact a similar legislation.

All of the above legislations have one thing in common. While they all permit rights of ownership over extracted resources, they do not yet assert any right of sovereignty over the exploited territory. This leads to a curious predicament in legal terms as a question naturally arises as to the presence of a distinction in the ownership over the territory and ownership over the extracted resources. The ideal answer to the above question would be arrived at by way of a consensus among States in a multilateral forum. However, given that such an approach has not been adopted yet, in the interim, it is proposed that the present legal framework be examined for the purposes of determining its sufficiency in arriving at an answer to the same.

As per Article I of the OST, exploration and use of outer space including the Moon and other celestial bodies is permitted. The term 'use' has been shown, in an earlier section of this article to include both economic and non-economic objectives, including the extraction of resources from outer space. Article II of the OST adds a precondition to such use as it states that regardless of any use, outer space is still not subject to national appropriation or claims of sovereignty. It is absolutely clear that this principle does not permit States to claim sovereignty rights over territory in outer space. By virtue of only States bearing international responsibility for private enterprises, which are to act under authorization by States, States can only permit what is not forbidden to itself.[56] This is in accordance with the principle of delegation in international law, where one can only delegate the authority that one possesses. It would be sanctimonious to established principles of international law, to say otherwise.

This is true even as per the common law system, where States need to have property rights first in order to confer or attribute them to its citizens.[57] This provision thus makes it clear that, even procedures as per customary international law that

[51] Nasa Office of Inspector General, 'Nasa's Management of Moon Rocks and Other Astromaterials Loaned For Research Education and Public Display: Report IG-12-007 Assignment No A-11-015-00' (2011) 5.

[52] Commercial Space Launch Competitiveness and Entrepreneurship Act (Spurring Private Aerospace Competitiveness and Entrepreneurship (SPACE) Act), (2015) Pub L No 114-90, 51303, 129 Stat 721.

[53] Hiroko Yotsumoto and Daiki Ishikawa, 'Japan' in Joanne Wheeler (ed.), *Space Law Review*, ed 1 (The Law Reviews 2019) 47.

[54] Setsulko Aoki, 'Domestic Legal Conditions for Space Activities in Asia' (2019) 113 AJIL UNBOUND 103, 104.

[55] Lucy Barnard, 'UAE to Finalise Space Laws Soon' (The National 7 March 2016) 1.

[56] Clarence W. Jenks, *Space Law* (Stevens and Sons 1965) 201.

[57] Wayne White, *The Legal Regime for Private Actors in Outer Space, in Space: The Free Market Frontier* (E.L. Hudgins 2002) 84.

permit the obtaining of sovereignty by States over territory on Earth, and subsequent allowances given by that State to its subjects over that territory, do not apply to outer space.[58]

The extension of the above-accepted restriction to the ownership over extracted resources, is however unclear. This is where legal scholarship divides, as there exist two major perspectives in this regard. First, that categorically refutes the right of any State, or private enterprise to claim ownership over extracted resources.[59] This seems to be premised on the notion that the OST prohibits any kind of exclusive rights and in fact designates outer space, including the Moon and other celestial bodies as an international common,[60] that is available for the use of all States, without discrimination.[61]

The other perspective carefully distinguishes between property rights over territory as being prohibited, and ownership rights over extracted resources as being permitted.[62] While the first perspective is clear, it is the second perspective that presents an interesting legal predicament. This position is premised on an understanding that the scope of the restrictions provided in Article II of the OST are vague and unclear.[63] This is even more so, given that there exists no specific *expressis verbis* prohibition of ownership of mined resources and any interpretation to that effect will only flow from logical implications of the provision.

According to one scholar who adopted this position, a workable distinction exists based on the foundational constraints of the OST, which can be taken as analogous to regimes governing other global commons, such as the United Nations Convention on the Law of the Sea.[64] This is premised on the fact that under the above regime, parties may establish some claim of ownership over extracted resources, and transfer the resources, despite being restricted from claims of sovereignty over the underlying land. However, despite adopting this stance, even this work recognizes that ownership even if permitted, is "undefined" and "attempts at resource extraction are bound to straddle the line between use and sovereign claims over land".[65] This statement

---

[58]Fabio Tronchetti, 'The Non-Appropriation Principle under Attack: Using Article II of the Outer Space Treaty in Its Defense', in *Proceedings of the 50th Colloquium on the Law of Outer Space* (Eleven International Publishing 2007) 526.

[59]Philip de Man, 'The Exploration of Outer Space and Celestial Bodies: A Functional Solution to the Natural Resource Challenge' 2010 Leuven Ctr. for Glob. Governance Studies, Working Paper No. 54, 13. https://ghum.kuleuven.be/ggs/publications/working_papers/2010/54DeMan. Accessed 6 September 2020.

[60]Iris Kyriazi, 'Outer Space as Global Commons' (Kataikos, 25 May 2020). https://katoikos.world/analysis/outer-space-as-global-commons.html. Accessed 28 August 2020.

[61]Treaty on Principles Governing the Activities of States in the Exploration and Use of Outer Space, Including the Moon and Other Celestial Bodies, (adopted 19 December 1966, entered into force 10 October 1967) 610 UNTS 205 (Outer Space Treaty) art 1.

[62]John G. Wrench, 'Non-Appropriation, No Problem: The Outer Space Treaty Is Ready for Asteroid Mining' (2019) 51 CASE W RES J INTL L 437.

[63]Kevin MacWhorter, 'Sustainable Mining: Incentivizing Asteroid Mining' (2016) 40 WM and MARY ENVTL L and POLY REV 645, 661.

[64]See footnote 62.

[65]See footnote 62.

indicates the prevailing uncertainty on the subject despite several legal interpretations and theory, which can only be laid to rest after a clear consensus is reached by the international community as a whole.

13.4 Conclusion and Recommendations

There is an immediate necessity to answer crucial legal questions such as, the scope of the freedom of 'use' and its relationship with the limitation of non-appropriation prior to establishing human settlements on the Red Planet, as such a trajectory would inevitably require the exploitation of resources available on the planet. This urgency flows from two primary reasons. First, that states and private enterprises interested in exploitation activities, will not be assuredly incentivized to invest and carry out such programs, if there exists no clarity with regard to their rights in outer space while engaging in such activities. This is so, considering that, any adverse declaration at a later stage, would spell billions in cost for them, to say the least. The second reason is that, if such a consensus is not arrived at soon, interested nations, particularly space faring nations, would continue to take the initiative and set out domestic legislation recognizing their rights to appropriate and own resources extracted from the Moon or even Mars, as the case may be. Their space programmes are not going to wait indefinitely for the conclusion of a proper regime in this regard, as it is hardly practical or economic for them to. This practical reality poses a significant disadvantage for the nations that lack space capabilities, as domestic actions by space faring nations may result in unequal territorial occupations, even if established only for the purpose of resource extraction and similarly unequal claims of ownership over extracted resources, leaving few, if any, resources for the non-space faring nations.[66] Therefore, the right of ownership over resources in outer space if any, along with the regime governing it, must be set out formally at the earliest possible.

The present framework of the OST attempts to provide for developments in space while simultaneously protecting the rights and future abilities of non-space faring nations. It does this by placing limitations in the nature discussed above, on the permitted activities of space faring nations. Although these limitations are for the greater purpose of promoting international cooperation in outer space activities among states,[67] it is unlikely that this objective will be realized without pivotal discussions being carried out immediately on the applicability and scope of several limitations contained in the OST.

It is recommended that multilateral efforts be made in this regard, at the UNCOPOUS, a proven space from which five major space treaties arose in the past decades. Even today, it remains the most appropriate forum to discuss and negotiate

[66] Abigail D. Pershing, 'Interpreting the Outer Space Treaty's Non-Appropriation Principle: Customary International Law from 1967 to Today' (2019) 44 YALE J INTL L 149, 150.

[67] Fabio Tronchetti, *The Exploitation of Natural Resources of the Moon and other Celestial Bodies*, vol 4 (Brill 2009) 63.

towards a consensus on the pressing questions relating to outer space that confront us today. There exists no need to render void the existing OST, but rather, a need to develop upon and clarify the fundamental principles expounded in the OST, as these principles, in their essence are sound and will continue to guide space development away from chaos. It is suggested that this be undertaken by developing an additional protocol that extends upon the freedoms and limitations contained in the mother treaty, and a separate treaty or set of regulations be created for the specific purpose of outlining and guiding human settlements on other celestial bodies. The additional protocol would be a viable path as it would provide a choice to present State parties to the OST to be bound by its entailments, and would therefore not impose the same. An additional treaty or regulations to regulate human establishments in outer space, should ideally not be limited to only Mars, but must instead be a compilation of the general regime and principles applicable to all future settlements on celestial bodies. This is suggested as a lesson from the OST, as it is difficult to anticipate and accordingly provide for all future advancements in aerospace technology and human aspirations. The sooner this consensus is arrived at and more comprehensive the framework, the clearer the horizon will be, for all of humankind's forays into the realms of outer space.

Radhey Soundarya Gnanesh is pursuing legal studies at Gujarat National Law University (India). She undertook a research internship in space law with the Indian Space Research Organisation (ISRO) at its Headquarters in Bangalore, India. She has also represented her University at the Leiden Sarin International Air Law Moot.

Chapter 14
Protecting the Million-Year Picnic: The Importance of Importing the Rule of Law to Mars

Yuk Chi Chan

Abstract The importance of ensuring that the Rule of Law becomes one of the first imports we bring onto the Red Planet cannot be understated. The first people on Mars will arrive on a commercial rocket, and the first settlement there will likely be funded by corporations, not governments. These private entities, though exempt from Rule of Law principles on Earth, will become the de facto authority on Mars, and they will have significant power over any colonists there. With such an enormous potential for tyranny, we should insist that corporate governors be obliged to uphold the same duties and obligations owed by national governments here on Earth, in the interest of improving the odds of success of the mission, preserving and protecting the individual rights and dignities of humanity's first ambassadors to Mars, and ensuring that our first steps out into a wider universe are guided by the very best of our legal traditions.

14.1 Introduction

> "'I've always wanted to see a Martian,' said Michael. "Where are they, Dad? You promised." "There they are," said Dad, and he shifted Michael on his shoulder and pointed straight down. The Martians were there. Timothy began to shiver. The Martians were there - in the canal - reflected in the water. Timothy and Michael and Robert and Mom and Dad. The Martians stared back up at them for a long, long silent time from the rippling water." Ray Bradbury[1]

These days, it is an inexorable consequence of any discussion on Mars that the conversation will turn to the question of to whom the task of settling the Red Planet will fall. Heavy is the question upon which rests all of humanity's hopes for becoming a multi-planetary species, but what it makes up for in heft, it lacks in mystery.

[1]Ray Bradbury, 'The Million-Year Picnic' in *The Martian Chronicles* (first published 1950, Simon and Schuster 2012) at [241].

Y. C. Chan (✉)
Anchor Orbital Systems, Harwell, Oxfordshire, UK
e-mail: yukchi@anchororbital.com

Donaldson & Burkinshaw LLP, Singapore, Singapore

A. Froehlich (ed.), *Assessing a Mars Agreement Including Human Settlements*, Studies in Space Policy 30, https://doi.org/10.1007/978-3-030-65013-1_14

It is uncontroversial to presume that private corporations, not governments, will be the first to arrive there. They will be the first to fly, first to establish a permanent habitat, and in all likelihood, the first to create a new Martian society.[2] The idea of a private colonisation of Mars has been around for a long time and seems all the more prescient today.[3]

It should also be uncontroversial to state that whilst on Mars, international space law will continue to apply.[4] The ship flown will still be a spacecraft, and the colonists will still be astronauts.[5] They will still retain citizenship from their respective countries, and their activities will still fall under the common definitions accepted in space law,[6] even if they never see *Terra* again. That having been said, the first Martians will be almost entirely beholden to their corporate governors for every resource they need to survive.[7] Food, shelter, equipment, even oxygen—there will be no opportunity to 'rough it' on Mars. Never before has there been such an enormous degree of dependence in any sort of settlement project, and consequently never has there been a greater opportunity for tyranny and exploitation.[8]

With the immense distance between Earth and Mars, what will the mere existence or letter of the law alone mean? What safeguards will there be to prevent a descent into an autocracy, underpinned by the will of a corporate sovereign?[9] And perhaps most importantly, what will be the legacy inherited by future colonists, ostensibly the first-ever biological beings born on Mars? One of exploitation and subjugation, or liberty and justice?

"It is the ideal of legality together with the Rule of Law that stands between us and a disempowering techno-managed future."[10] Of all the things we can bring with us to Mars, perhaps most important of all is the Rule of Law: the principle that all persons

---

[2]This is, at least, the prevailing sentiment among commentators. See: Michael Solana, 'In Mars Co. We Trust' (*Quillette*, 27 December 2019). https://quillette.com/2019/12/27/in-mars-co-we-trust-understanding-the-coming-interstellar-corporatocracy/. Accessed 31 August 2020.

[3]Peter Ward, *The Consequential Frontier* (Melville House 2019) at [178–180].

[4]Generally, each of the substantive articles of the Outer Space Treaty extends to Mars as a celestial body. See: Dr. Dionysia-Theodora Avgerinopoulou and Katerina Stolis, 'Current Trends and Challenges in International Space Law' (53rd European Space Sciences Committee Plenary Meeting, Athens, 1 July 2017); Karl Lieb, 'State Sovereignty in Space: Current Models and Possible Futures' (2015) 13(1) *Astropolitics* 1.

[5]Or at the very least some similar category of spacefarer. See Stephan Hobe, 'Space Tourism as a challenge to the Astronaut Concept' in Gabriel Lafferranderie and Sergio Marchisio (eds.), *The Astronauts and Rescue Agreement: Lessons Learned* (European Centre for Space Law 2011) at [71].

[6]Broadly, this refers to both the major space treaties, as well as the assorted international guidelines, national legislation, and other 'soft law' measures that have been promulgated throughout the years. See generally: Francis Lyall and Paul Larsen, *Space Law: A Treatise* (2nd Edn., Routledge 2018); Irmgard Marboe (ed.), *Soft Law in Outer Space* (Böhlau Verlag 2012).

[7]Charles Cockell, *Extra-terrestrial Liberty* (Shoving Leopard 2013).

[8]Erik Persson, 'Citizens of Mars Ltd.' in Charles Cockell (ed.), *Human Governance Beyond Earth* (Springer 2015) at [125].

[9]John Austin, *The Province of Jurisprudence Determined* (first published 1832, CUP 1995).

[10]Roger-Brownsword, 'Technological management and the rule-of-law' (2016) 8 *L.I.T.* 100 at [138].

and authorities, public or private, are equally bound by and entitled to the benefit of laws whose effects are antecedent to their public declarations, and which are openly administered by impartial tribunals.[11] Importing it will cost no extra fuel, nor will it consume any resources upon arrival, and yet it will have a material impact on the lives and liberties of the first colonists as well as subsequent generations of Martians by ensuring that their individual rights and responsibilities are not subordinated to mere ends.[12]

Much of the foregoing discussion on the subject has been focused on specific practical and legal considerations.[13] This essay is instead about the broader discussion of the sort of normative shape that Martian law should take. The imposition of robust Rule of Law principles[14] will not be the most efficient means of running a Mars colony;[15] they are barely efficient even here. Efficiency is not its intended objective. The purpose of the Rule of Law, dogmatic and axiomatic as it is,[16] is to constrain power so that those who labour under its watch may do so freely and peaceably. To that end, it is the best tool for the job. And on that lonely red ball of dust, so far away that even radio signals take at least five minutes to arrive from Earth, such a guarantee should be more prized, not less.

14.2 Space Law on Mars

Before considering the necessity and application of the Rule of Law to a Martian settlement, it is first important to explain how existing space law would initially apply.

The modern regime of space law comprises a mix of public international law instruments and corresponding domestic legislation and regulation.[17] Mars falls under the ambit of the 1967 Outer Space Treaty (OST)[18] as a celestial body, and is

---

[11]Tom Bingham, *The Rule of Law* (Penguin 2010) at [8].

[12]Ronald Dworkin, *Law's Empire* (Hart 1986) at [93].

[13]Sara Bruhns and Jacob Haqq-Misra, 'A pragmatic approach to sovereignty on Mars' (2016) 38 *Space Policy* 57.

[14]Paul Craig, 'Formal and Substantive Conceptions of the Rule of Law' (1997) *Pub.L.* 467.

[15]At any rate, Elon Musk's goal of Martian direct democracy doesn't seem to be aimed at efficiency either. See: Loren Grush, 'Elon Musk thinks he best government for Mars is a direct democracy' (*The Verge*, 2 June 2016). https://www.theverge.com/2016/6/2/11837590/elon-musk-mars-government-direct-democracy-law-code-conference. Accessed 5 September 2016.

[16]Alain Supiot, *Homo Juridicus: On the Anthropological Function of the Law* (Saskia Brown tr., Verso 2007) at [xv].

[17]Joanne Gabrynowicz, 'Space Law: Its Cold War Origins and Challenges in the Era of Globalization' (2004) 37 *Suffolk U.L.Rev* 1041.

[18]Treaty on principles governing the activities of States in the exploration and use of outer space, including the moon and other celestial bodies (adopted 27 January 1967, in force 10 October 1967) (1968) 610 UNTS 205.

governed by the various principles contained therein like non-appropriation and non-militarisation, as well as subsequent instruments concerning liability and registration. By virtue of Article VI of the OST, national governments are ultimately responsible for all the activities of non-governmental entities, including private companies. This, in conjunction with the liability and attribution stipulations in Articles VII and VIII, provides state parties with jurisdiction and control over any space objects launched and registered under their authority,[19] including any corporate Mars colonies.

However, within the larger sphere of space law exists a number of unresolved issues[20] that readily lend themselves to mischief within the context of a Mars colony.[21] And yet, aside from the poorly-subscribed Moon Agreement[22] and the still-untested Artemis Accords,[23] there is still little clarification offered by the existing legal regime to aid in navigating the specific questions that accompany the prospect of establishing a long-term base on the Red Planet.[24]

For example, what would be the applicability of domestic space legislation, such as the US SPACE Act of 2015[25] of which Title IV in effect allows US citizens to lay claim to harvested extra-terrestrial resources,[26] within the context of a Martian settlement? Assuming the colony adopts a similar arrangement to the International Space Station wherein each colonist continues to be subject to their home country's national laws,[27] it is conceivable that colonists from other countries without similar enabling legislation like the SPACE Act would be unable to lawfully lay claim to

[19] Rada Popova and Volker Schaus, 'The Legal Framework for Space Debris Remediation as a Tool for Sustainability in Outer Space' (2018) 5(2) *Aerospace* 55 at [64].

[20] For example, the matter of space debris. See: Kimitake Nakamura, 'Space Debris and Liability Schemes under International Law' (2020) 95(3) *Waseda Law Review* (translated and available at. https://spi.elliott.gwu.edu/2020/06/10/space-debris-and-liability-schemes-under-international-law/). Accessed 2 September 2020; Paul Larsen, 'Solving the Space Debris Crisis' (2018) 83(3) *JALC* 475.

[21] Frans von der Dunk, 'International Space Law', in Frans von der Dunk (ed.), *Handbook of Space Law* (Edward Elgar 2015) at [53–4].

[22] Agreement Governing the Activities of States on the Moon and Other Celestial Bodies, UN Doc. A/34/664 Nov. 1979 (adopted 5 December 1979, in force 11 July 1984) 1363 UNTS 3.

[23] NASA, 'The Artemis Accords'. https://www.nasa.gov/specials/artemis-accords/index.html. Accessed 2 September 2020.

[24] Caroline Haskins, 'The Legal Battle to Colonize Mars' (*The Outline*, 15 March 2018). https://theoutline.com/post/3739/mars-colony-settlement-spacex-elon-musk-trump?zd = 1&zi = hns3qj4x. Accessed 5 September 2020.

[25] Commercial Space Launch Competitiveness Act, Pub. L. No. 114-90, 129 Stat. 704.

[26] Frans von der Dunk, 'The US Space Launch Competitiveness Act of 2015' (2015) 98 *Space, Cyber, and Telecommunications Law Program Faculty Publications*.

[27] Gabriella Catalano Sgrosso, 'Legal Status of the Crew in the International Space Station' (1999) 42 *Proceedings of the International Institute of Space Law* 35; Lyall and Larson (n. 6) at [129].

harvested resources, creating an immediate disparity in legal rights. This incongruence also exists where there are variations in laws between countries with comprehensive systems of space legislation, as well as between countries where some have developed space legislation and others have not.[28]

This does not even begin to cover the various other inadequacies, such as the lack of any means to effectively regulate private entities short of simply delegating the responsibility to State parties. Lack of enforceability is a pertinent issue that weighs upon the legitimacy of any proposed system.[29] Purely private international law-based regimes,[30] presumably backed by some means of dispute resolution such as arbitration,[31] would be untenable within the context of a settlement because they would only regulate the behaviours of corporate entities amongst each other. Human beings are not corporations, and lack many of the motivations and traits upon which private international legal regimes are predicated in order to function.[32] Lacking any sort of normative basis or objective beyond efficiency, such regimes are hence wholly inadequate because they fail to provide for what is arguably the most important element of a colony—the colonists themselves.

Finally, it is worth noting that these quibbles pertain to administrative considerations for those still on Earth, but fail to touch on the reality of life on the Martian surface itself.[33] International space law has created a general, macro-level framework towards space that provides for none of the finer fundamentals necessary for a functioning legal system. There are no mechanisms for dispute resolution, decision-making, or environmental protection. The prospect of exporting our terrestrial law to space has rendered each aspect of it open to re-examination,[34] and many practically ancient concepts like property rights[35] have now become 'essentially contested'.[36]

In light of all of this, it would seem that the present space law regime is inadequate in governing a Mars colony. A separate agreement, or perhaps an entirely new legal system, will have to be created to rise to the challenge. That is where the Rule of Law must come in.

---

[28]Frans von der Dunk, 'Billion-dollar questions? Legal aspects of commercial space activities' (2018) 23 *Unif.L.Rev.* 418 at [445].

[29]For example, as in the Australian case of *Humane Society International v Kyodo Senpaku* [2005] FCA 66, wherein the Australian Court's ruling remains unenforced.

[30]Alexander Salter, 'Ordering the Cosmos: Private Law and Celestial Property Rights' (2017) 82(2) *JALC* 311.

[31]PCA Optional Rules for Arbitration of Disputes Relating to Outer Space Activities (PCA Rules on Outer Space Disputes).

[32]Salter (n. 30) at [322–328].

[33]Igor Levchenko et al., 'Mars Colonization: Beyond Getting There' (2019) 3 *Global Challenges* 1800062.

[34]Senjuti Mallick and Rajeswani Rajagopalan, 'An Examination of the Potential of Space Mining and its Legal Implications' (ORF Occasional Paper No. 182, Observer Research Foundation, January 2019) at [6–13].

[35]John Locke, *Second Treatise on Government* (first published 1689, Everyman 1993).

[36]Walter Gallie, 'Essentially Contested Concepts' in Max Black (ed.), *The Importance of Language* (Cornell University Press 1962) at [121].

14.3 Martian Law

Mars is a lethal environment, lacking in virtually everything necessary to sustain life. Upon arrival, colonists will be entirely dependent on their corporate governors for survival: supplies, shelter, and even oxygen. In turn, this unprecedented degree of control engenders the possibility of tyranny and autocratic oppression on a scale never before seen.[37] As long as the arm of the law may be, hundreds of millions of miles may prove to be too far for the current, arguably dysfunctional and inadequate, system of international space law to be capable of effecting any guidance, both normative and practical.

Hence it behoves us today to consider the creation of a local Martian legal system which takes into account the unique challenges of a colony and the measures necessary to address them.[38] Whatever jurisprudential philosophy is chosen to underpin this legal system will have to be sensitive to the needs and objectives of both the colonists and the corporation in charge of the colony;[39] beyond that however, the additional application of the Rule of Law is critical in light of the obvious opportunities for administrative abuse and exploitation, as well as in the interest of the long-term stability of the colony itself.

14.3.1 The Rule of Law

The Rule of Law is most easily understood according to its core tenets[40]: transparency and accountability, predictability and consistency, and equality before the law.[41]

Accountability holds that all actors are subject to law's reign, and that "be you ever so high, the law is above you".[42] Transparency in turn enables accountability by shining light into otherwise opaque corners and ensuring that nothing escapes justiciability.[43] Predictability underpins several important principles:[44] non-retroactivity,

---

[37]Charles Cockell, 'Extraterrestrial Liberty' in Charles Cockell (ed.), *Human Governance Beyond Earth* (Springer 2015) at [24–32].

[38]Tony Milligan, 'Rawlsian Deliberation About Space Settlement' in Charles Cockell (ed.), *Human Governance Beyond Earth* (Springer 2015) at [17–22].

[39]Mukesh Bhatt, 'Constituting Outer Space' in Charles Cockell (ed.), *Human Governance Beyond Earth* (Springer 2015) at [160].

[40]Brian Tamahana, *On the Rule of Law: History, Politics, Theory* (CUP 2004); Paul Gowder, *The Rule of Law in the Real World* (CUP 2016) at [3]; Jeremy Waldron, 'The Concept and the Rule of Law' (2008) 43 *Ga.L.Rev.* 1.

[41]Monika Zalnieriute et al., 'The Rule of Law and Automation of Government Decision-Making' (2019) 82(3) *MLR* 425 at [429].

[42]*Gouriet v Union of Post Office Workers* [1977] 1 QB 729 at [762] (Denning M.R.).

[43]Lyria Moses and Louis de Koker, 'Open Secrets: Balancing Operational Secrecy and Transparency in the Collection and Use of Data for National Security and Law Enforcement Agencies' (2017) 41 *MULR* 530 at [534–537].

[44]Bingham (n. 11) at [54].

clarity, non-arbitrariness, and the judicial practice of adherence to precedent.[45] Equality before the law means treating like cases alike[46] and ensures that one's particulars or peculiarities does not bind one under the law any differently from another.[47]

But the Rule of Law is a great deal more than just the sum of its parts.

It is an ideal, built upon principles that aspire toward a particular form of society—one that is commonly accepted to be most happy because of the traits it bears, and the protections it provides for the rights that it confers.[48]

The practical importance of the Rule of Law as a prerequisite of good governance is universally agreed upon.[49] Predictability and consistency enable long-term planning,[50] whilst accountability and equality ensure the visibility of justice, making unfavourable decisions more palatable to those inconvenienced, their personal disappointment notwithstanding.[51] It also trammels the exercise of power, preventing those with authority from doing as they wish with impunity.[52] These conditions in turn provide the stability necessary to allow societies to grow, and civilisations to thrive.[53]

The Rule of Law also serves an important normative function. It informs not just what is, but what ought to be, and it imposes obligations and confers rights even in the absence of congruent practices or in the face of contrary ones. These rights and obligations exist because they are "correct principles of morality, not, *ex hypothesi*, because [they correspond] to an accepted practice."[54] Certain rights, like freedom of speech or due process,[55] are timeless and almost invariably desirable, such that despite whatever problems Mars presents, they will be worth fighting for because they lend credence to otherwise empty rhetoric like that of 'self-evident truths,' 'simple and incontestable principles,' or even the 'highest aspirations of the common people'.[56]

---

[45]Jeremy Waldron, 'Stare Decisis and the Rule of Law' (2012) 111 *Mich.L.Rev.* 1.

[46]Peter Westen, 'The Empty Idea of Equality' (1982) 95 *Harv.L.Rev.* 537 at [542–548].

[47]Yoav Dotan, 'Why Administrators should be Bound by their Policies' (1997) 17 *OJLS* 23 at [23].

[48]A.V. Dicey, *An Introduction to the Study of the Law of the Constitution* (First published 1885; 9th edn., Macmillan, 1945) at [195].

[49]Bingham (n. 11) at [171–174].

[50]*R v Gul* [2013] UKSC 64 at [36] (Neuberger LJ).

[51]Michel Rosenfeld, 'Substantive Equality and Equal Opportunity' (1986) 74 *CLR* 1687 at [1698].

[52]*Railway Express Agency Inc. v New York* [1949] 336 US 106 at [112–113] (Jackson J).

[53]Tim Cowan, '"Justice Delayed is Justice Denied": The Rule of Law, Economic Development and the Future of the European Community Courts' 2008 4(1) *European Competition Law Journal* 1 at [3–5].

[54]Jules Coleman, 'Negative and Positive Positivism' (1982) 11(1) *J.Leg.Stud.* 139 at [140].

[55]H. L. A. Hart, 'Positivism and the Separation of Law and Morals' (1958) 71 *Harv.L.Rev.* 593 at [623–624].

[56]Charles Cockell, 'Human Governance and Liberty Beyond Earth' in Charles Cockell (ed.), *Human Governance Beyond Earth* (Springer 2015) at [3].

14.3.2 Why the Rule of Law

This article presupposes that colonisation of Mars will take place, either out of necessity or our collective ambition as a species.[57] It will be difficult, and dangerous to say the least. It will take humanity's best, brightest, and bravest to even stand a chance at succeeding.

Extreme isolation, lethal environments, and countless technical challenges already face our intrepid colonists. It is difficult to imagine any of them also consenting to the further possibility of being subjected to arbitrary rule[58] or being subjugated into de facto corporate slavery. The Rule of Law, as an institutional safeguard against this, can therefore become a powerful recruiting tool, or at the very least prevent would-be settlers from being deterred. It is no secret that corporations are not driven by altruism,[59] and given the immense power disparity between colonists and their governors, a formal and highly visible guarantee against the abrogation of rights or other abuses of power will go a long way toward assuaging fears and instilling confidence. The Rule of Law, by guarding against the possibility of minority oppression,[60] will also encourage a greater diversity of backgrounds and skillsets amongst colonists. It is only logical that we as a planet would want our first ambassadors into the cosmos to comprise our best from all walks of life.

The Rule of Law will also ensure the stability, and hence long-term survivability, of the colony itself. Science fiction is awash with stories of despotic interstellar governments that incite their own downfalls through the fomenting of popular rebellion.[61] In reality however, with virtually zero margin for error, any sort of discontent in a Martian colony will likely prove fatal. Mars itself will provide innumerable opportunities for unhappy compromises and unpleasant choices. Without the public guarantees provided by the Rule of Law, the legitimacy of each unpopular decision will be suspect. Conversely, confidence in the fairness of a system of governance can lead people to tolerate even the most heavy-handed of restrictions to their personal liberties—a spirit well-suited to the harsh realities of a Martian settlement.[62] On Earth, draconian counter-terrorism and security regimes are endured and even supported, so long as those subject to it believe that said regimes are just and principled.[63] The Rule of Law thus serves an important role as both a visible reminder and an assurance of

---

[57] Derek Thompson, 'Is Colonizing Mars the Most Important Project in Human History?' (*The Atlantic*, 29 June 2018). https://www.theatlantic.com/technology/archive/2018/06/could-colonizing-mars-be-the-most-important-project-in-human-history/564041/. Accessed 31 August 2020.

[58] R. George Wright, 'Arbitrariness: Why the Most Important Idea in Administrative Law Can't Be Defined, and What This Means for the Law in General' (2010) 44 *U.Rich.L.Rev.* 839.

[59] Persson (n. 8) at [126–130].

[60] *Railway Express* (n. 52).

[61] Stephen Baxter, 'The Birth of a New Republic: Depictions of the Governance of a Free Moon in Science Fiction' in Charles Cockell (ed.), *Human Governance Beyond Earth* (Springer 2015).

[62] Adam Stevens, 'The Price of Air' in Charles Cockell (ed.), *Human Governance Beyond Earth* (Springer 2015) at [59].

[63] Aziz Huq et al., 'Mechanisms for Eliciting Cooperation in Counterterrorism Policing' (2011) 8 *JELS* 728 at [760].

justice,[64] placating individual suspicions and distrust in order to enable cooperation in furtherance of survival.

Corollary to the above, a Martian colony is a major project and will require a substantial amount of funding.[65] Investors, both public and private, are generally rational actors, and it stands to reason that they would seek to limit the possibility of the colony failing as a result of a descent into chaos, considering the enormous economic loss such an event would occasion. On Earth, the Rule of Law makes investments in certain countries more attractive by ensuring certain conditions conducive to desired outcomes, whilst guaranteeing against negative ones.[66] Similarly on Mars, through enabling societal stability and enhancing cooperation, it will materially improve the chances of the colony's success, and hence serves as a convenient risk mitigation strategy. The imposition of Rule of Law principles will thus begin paying dividends even before the first colony ship is launched by helping to attract the funding necessary to make the mission a reality.

Finally, it is important to consider the legacy being left for posterity. Future generations of humans born on Mars will have no memory of life on Earth, no understanding of our culture or traditions aside from what their parents brought with them. Stories like the Magna Carta and the Emancipation Proclamation will be as distant to them as the Earth itself. And yet, we are each born with an innate sense of justice: we are rational, and understand virtues like fairness, equality, and liberty.[67] The Rule of Law represents the distillation of over a thousand years' of thought and philosophy,[68] and while it is far from being a settled concept, there is still some wisdom to be found in it. What sense is there in depriving future Martians of this wisdom by forcing them to begin again with a *tabula rasa*? Worse still, considering the normative nature of the Rule of Law, what message are we conveying about its values and the rights it protects by choosing to leave it behind? From an ideological standpoint, bringing the Rule of Law to Mars represents more than simply a commitment to its principles; it is also an indicator that we, as a species, still believe that its principles are worth committing to.

14.3.3 Adapting the Rule of Law

The importance of importing the Rule of Law to Mars having now been established, we turn to the question of how best to adapt its application to the circumstances. Traditionally, it has only been imposed on national governments in view of the power

---

[64] *R v Sussex Justices ex parte McCarthy* [1924] KB 256 at [259] (Lord Heward CJ).

[65] Jacob Haqq-Misra, 'Towards a Sustainable Land Use Policy for Mars' in Charles Cockell (ed.), *Human Governance Beyond Earth* (Springer 2015).

[66] Mark Tushnet, 'Authoritarian Constitutionalism' (2015) 100 *Cornell L.Rev.* 391 at [424].

[67] Supiot (n. 16) at [4–20].

[68] Bingham (n. 11) at [10].

disparity between each individual and the state they belong to.[69] Its application is, therefore, not predicated on the nature of the entity it is being applied to, but rather the dynamic between the authority and the subject. A ready analogy for the extension of Rule of Law principles to private entities exists in 'Big Tech' companies like Facebook and Twitter, which have developed an enormous amount of regulatory power as a result of the ubiquity of social media sites and their popularity as platforms for the exercise of free speech.[70]

Given that the corporate government of a Martian colony will in fact be a government, the immediate applicability of the Rule of Law is far more apparent. Much as with tech companies, these private administrators will be able to effect control over human behaviour and enforce desired changes through design and active intervention. Whatever rules they create, even if they lack the hallmarks of legislation, will nonetheless regulatory in nature;[71] in fact it is precisely because they lack these hallmarks that they warrant the extension of these principles.

In adapting the Rule of Law, it is also important to understand its purpose. It is not prescriptive in and of itself, but rather informs the form and function of promulgated rules. It is an attitude with which problems are meant to be approached.

Therefore, there must first be some means of contesting rules and appealing decisions. It is impossible to prescribe legislation for every possible outcome, and inevitably, something will be missed.[72] Having a mechanism through which these mistakes can be reviewed and corrected is key to maintaining both the enforceability of rights and the ability of the colony to quickly adapt to emergent circumstances.[73]

Second, the law must serve a purpose and ultimately be of some benefit.[74] The Rule of Law imposes constraints for the betterment of society; without said betterment, a legal system cannot claim its legitimacy, and hence cannot be said to be law, let alone embody the Rule of Law.[75] The law is not intended to be an exercise in futility. It should guide conduct, and in so doing, provide reasons that we as rational beings can comprehend, internalise, and thus be better off for it.[76]

Lastly, as the embodiment of our proudest values, the Rule of Law should confer an integrity to the colony's system of rules that necessitates their interpretation according

---

[69] Richard Mulgan, *Holding Power to Account: Accountability in Modern Democracies* (Palgrave Macmillan 2003).

[70] Stuart Macdonald et al., 'Regulating terrorist content on social media: automation and the rule of law' (2019) 15(2) *Int. J.L.C.* 183.

[71] Roger Brownsword, 'Lost in translation: legality, regulatory margins, and technological management' (2011) 26 *Berkeley Tech.L.J.* 1321 at [1327].

[72] Mireille Hildebrandt, 'Law as computation in the era of artificial legal intelligence' (2018) 68 *U.T.L.J.* 12 at [34].

[73] Brian Sheppard, 'Warming up to inscrutability: How technology could challenge our concept of law' (2018) 68 *U.T.L.J.* 36 at [54].

[74] Brian Leiter, *Naturalizing Jurisprudence* (OUP 2007) at [172].

[75] Joseph Raz, *The Authority of Law* (Clarendon, 1979) at [v-vi].

[76] Michael Sevel, 'The Constitution of Authority' (2014) 5(2) *Jurisprudence* 430; Joseph Raz, 'The Problem of Authority: Revisiting the Service Conception' (2006) 90 *Minn L.Rev.* 1003.

to their best possible light.[77] Whenever there are doubts, the fairer and more just interpretation should prevail. "The end of law is not to abolish or restrain but to preserve and enlarge freedom."[78] The attitude adopted towards adjudication and governance should thus reflect the very highest and most honourable of our legal traditions.

14.4 Conclusion

We ought not to be cavalier about our first step out into a larger universe, or console ourselves that 'we'll figure out that stuff later'. We must consider these questions today, whilst time is still on our side, so that we will know better than to defer to convenient appeals tomorrow. This chapter has shown that our current international space law regime is inadequate by itself to govern a Martian settlement. More will have to be done, and when the time comes, our choice to import the Rule of Law will institute structures that facilitate long-term social stability, and ultimately improve the colony's chances of survival.

There will be many difficult choices ahead, hardships to be endured, and sacrifices to be made. We as a species have our work cut out for us. But we should take heart that even now, we stand on the shoulders of giants. As we look up, look up! and let loose cries of "Ad Astra", we should remember to take with us the hard-won legal wisdom of which we are the inheritors. Mars may be a new beginning, but that does not mean we need to start from scratch.

Chan Yuk Chi is a LL.B. graduate from Durham University Law School, and the Co-Founder and General Counsel of Anchor Orbital Systems. His areas of interest within space law include orbital debris mitigation and small satellite regulation. He is also interested in AI law and public policy, constitutional law, and jurisprudence.

[77]Dworkin (n. 12) at [218].

[78]Locke (n. 35) at [142].

Chapter 15
From Antarctica to Mars: Developing a Mars Treaty System

John Tziouras

Abstract Under current international space law, there has yet to be a definite answer about the legal status of a possible human settlement on Mars. Sending humans into deep space remains a long-term endeavour, but in the meantime it is important to continue studying the legal aspects of the first human settlement on a celestial body. A few pioneering industries have already been preparing the technology for this eventuality; when technological and financial hurdles have been overcome, a disruptive new international regime would evolve. This article examines some of the legal issues associated with the first settlement on Mars by focusing on the lessons that can be learned from the governance of a region that is facing similar challenges: Antarctica. Thus, the Antarctic Treaty System could serve as a reference for drafting a Mars Agreement.

15.1 Introduction

The concept of human settlements on Mars presents numerous political, moral and security challenges, far from technological or health hurdles. From 'Mars Direct'[1] plans for Mars colonisation in the 1990s to A. Weir's novel *The Martian*,[2] there is a giant leap to be made for the first settlement.

Recent developments with SpaceX's *Starship* make it one of the most ambitious test programmes in rocket history,[3] in keeping with the company's goal to send

[1] Robert Zubrin, *The Case for Mars: The Plan to Settle the Red Planet and Why We Must* (Touchstone 1997).

[2] Andy Weir, *The Martian: A Novel* (1st edn, Crown Publishers 2014).

[3] Chris Bergin, 'Starship family grows ahead of new test phase' (*Nasa spaceflight*, 20 September 2020). https://www.nasaspaceflight.com/2020/09/starship-family-new-test-phase/. (All website cited in this article were last accessed and verified on September 30, 2020).

J. Tziouras (✉)
Faculty of Law, Aristotle University of Thessaloniki, Thessaloniki, Greece
e-mail: tziouras@law.auth.gr

A. Froehlich (ed.), *Assessing a Mars Agreement Including Human Settlements*, Studies in Space Policy 30, https://doi.org/10.1007/978-3-030-65013-1_15

humans to Mars.[4] However, for NASA, the first 'small step' for an interplanetary human journey must first pass through the Moon. Under the *Artemis* programme, National Aeronautics and Space Administration's (NASA)'s human lunar exploration plans would lay the groundwork for sending astronauts to Mars.[5]

However, there is a difference between a permanent settlement on planetary surfaces and an advanced human mission to Mars. As a result, the broadly used term, 'space colonisation', is somewhat inaccurate. This is not only because it usually refers to 'permanent and self-sufficient human habitats beyond Earth' and because 'the main goal of the planned missions is not subjugation, but exploration', but also because these activities should be distinguished from the negative history of colonisation on Earth in order to develop a common, mutually accepted agreement.[6]

In attempting to disentangle the complexities of future urban settlements on the Moon and Mars, it would be preferable to adopt a step-by-step approach and draw on lessons from other legal regimes, such as the Antarctic Treaty System, the Law of the Sea, the International Telecommunication Union (ITU) model or the International Space Station Intergovernmental Agreement (ISS). Initially, this would enable the establishment of a possible multilateral regime for planetary settlement and resources activities, which could in turn help the drafting of a Mars Agreement at a later stage.

The present article's aim is to identify some of the lessons that can be learned from the governance of other 'un-owned' regions facing similar challenges. Antarctic governance provides a useful case study of cooperation, beginning with the adoption of the Antarctic Treaty and continuing with subsequent agreements.

15.2 Defining the Legal Framework: Applicable International Treaties and 'Controversial' Principles on Sovereignty and Jurisdiction

It is believed that Antarctica is the only continent without a history of human habitation; the same is believed to be true for Mars. However, this did not prevent states from claiming sovereignty over Antarctica. Although these claims are not so persistent and of a more historical nature[7]—as both Russia and the United States consider themselves pioneers of Antarctica and entitled to sovereignty over the territory under

---

[4]SpaceX, 'Mars & Beyond: The Road to Making Humanity Multiplanetary' (SpaceX). https://www.spacex.com/human-spaceflight/mars/.

[5]Erin Mahoney, 'Sending American Astronauts to Moon in 2024: NASA Accepts Challenge' (*NASA*, 9 April 2019). https://www.nasa.gov/feature/sending-american-astronauts-to-moon-in-2024-nasa-accepts-challenge/.

[6]Marko Kovic, 'Political, moral, and security challenges of space colonization' (2018) ZIPAR Discussion Paper Series. https://zipar.org/discussion-paper/political-moral-security-challenges-space-colonization/.

[7]'Thaw in International Law? Rights in Antarctica under the Law of Common Spaces' (1978) 87 The Yale Law Journal 804.

their *imperium*[8]—the prospect of accessing rich natural resources has the potential to revive old and new claims of other states.[9]

According to the theory of acquisition, a state may acquire territorial titles by means of occupation, sovereignty, annexation, accretion, prescription and cession.[10] However, the concept of 'spatiality' is applicable to all of these models, which means that spaces are not merely inert physical objects but rather dynamic products of human perception and representation.[11] Although this article does not adopt H. LeFebvre's key theoretical concept of territory,[12] it is important to highlight the recent shift in interpretation concerning the principles of non-appropriation and common heritage of mankind.

15.2.1 Who Owns Mars? The Debate Over 'Un-Owned' Regions, Sovereignty and Jurisdiction

Under the Outer Space Treaty (OST) of 1967, sovereignty is excluded under the terms of Article II. The majority of scholars support the view that Moon and other celestial bodies cannot become the territory of one sovereign state. According to this view, the Moon and other celestial bodies are not subject to national appropriation by claim of sovereignty, by means of occupation or by any other means.[13] Furthermore, in accordance with the OST, the Moon Agreement (Article 11, para 2) features the principle that the Moon shall not be subject to national appropriation by claims of sovereignty, by means of use or occupation or by any other means.[14] As a result, the Moon cannot be exclusively appropriated neither by the planting of national flags on its surface nor by its use for peaceful experiments.

The same interpretation can be made concerning Mars. According to some scholars, 'if Mars is a celestial body and if Mars is in outer space, then under the

[8]Yana Evgenyevna Brazovskaya and Gulnara Flurovna Ruchkina 'Current Human Impact on Antarctic Seabed Environment and International Law' in Catherine Banet (ed), *The Law of the Seabed: Access Uses and Protection of Seabed Resources* (Brill Nijhoff 2020).

[9]Shirley V. Scott, 'Ingenious and innocuous? Article IV of the Antarctic Treaty as imperialism' (2011) 1 The Polar Journal 51.

[10]Ogunsola O. Ogunbanwo, *International Law and Outer Space Activities* (Springer Science + Business Media 1975).

[11]Neil Brenner and Stuart Elden, 'Henri Lefebvre on State, Space, Territory' (2009) 3 International Political Sociology 353.

[12]John G. Wrench, 'Non-Appropriation, No Problem: The Outer Space Treaty Is Ready for Asteroid Mining' (2019) 51 Case Western Reserve Journal of International Law 437.

[13]Treaty on Principles Governing the Activities of States in the Exploration and Use of Outer Space, including the Moon and Other Celestial Bodies, adopted on 19 Dec. 1966, entered into force on 10 Oct. 1967, 610 UNTS 205.

[14]Agreement Governing the Activities of States on the Moon and Other Celestial Bodies, adopted on 5 Dec. 1979, entered into force on 11 July 1984, 1363 UNTS 3.

Article II of the OST, there can be no claims of sovereignty on the Red Planet'.[15] Thus, the principle of non-sovereignty can also be applied to Mars, even if there is no precise definition of the term 'celestial body'.[16]

Is that the only interpretation? Questions have begun to surface about whether the Moon is a *res nullius* or a *res communis*. In an older study, O. Ogunbanwo noted that 'under International Customary Law, outer space is a *res communis* while celestial bodies are *res nullius*, i.e. capable of being under national sovereignty'. According to this view, 'Article II of the OST has abrogated the *res nullius* notion of the International Customary Law', which underpins B. Cheng's view that state parties cannot exercise territorial jurisdiction over celestial bodies.[17]

However, according to another view, there are no obstacles to resources extraction in outer space, even when the non-appropriation principle is accepted as a restriction on sovereign claims to a celestial body. In addition, the concept of a sovereignty claim in such a context would be different from a jurisdictional claim. Jurisdiction is an aspect of state sovereignty and concerns the state's ability to exercise legal control over its territory and people. What Article II of the OST establishes is that concepts such as occupation and national appropriation under international law do not apply in outer space.[18] Under this interpretation, a research mission established on the surface of Mars by the United States—or a non-governmental entity registered and licensed by the United States—would not be regarded as a sovereign claim but rather a jurisdictional claim.

Questions are also starting to be raised about where sovereignty or national jurisdiction ends. For example, do any areas on Earth lie beyond the limits of national jurisdiction or sovereignty?

The answer can be found in the Law of the Sea, which regulates areas beyond national jurisdiction (e.g. some marine areas, the high seas and the seabed). As an umbrella treaty that covers all ocean uses, the Convention on the Law of the Sea (UNCLOS) was designed to serve as a core framework for a growing number of sectoral agreements that address one or more particular ocean uses. In areas that lie beyond national jurisdiction, specialised conventions cover several issues (e.g. international shipping, fisheries and waste disposal at sea).[19] However, these conventions do not have a binding effect on non-parties to the agreement, except as specifically

---

[15] Zach Miller, 'Space Settlement and the Celestial Subjectivity Model: Shifting Our Legal Perspectives of the Universe' in Annette Froehlich (ed), *A Fresh View on the Outer Space Treaty* (Springer International Publishing AG 2018).

[16] Tanja Masson-Zwaan and Mahulena Hofmann, *Introduction to Space Law* (4th edn, Wolters Kluwer 2019) 34.

[17] Ogunsola O. Ogunbanwo, *International Law and Outer Space Activities* (Springer Science + Business Media 1975).

[18] Frans von der Dunk, 'Sovereignty Versus Space- Public Law and Private Launch in the Asian Context' (2001) 5 Singapore Journal of International & Comparative Law 22.

[19] Lee A. Kimball, 'The International Legal Regime of the High Seas and the Seabed Beyond the Limits of National Jurisdiction and Options for Cooperation for the establishment of Marine Protected Areas in Marine Areas Beyond the Limits of National Jurisdiction'. Secretariat of the Convention on Biological Diversity, Montreal, Technical Series no. 19 (2005).

noted in international customary law. Similarly, the international seabed is beyond sovereignty.

Furthermore, there are three legal regimes that share the non-appropriation principle's ban on sovereign claims but at the same time permit the ownership of extracted resources. One of them is the Antarctica Treaty System (ATS).[20] Sovereignty claims related to Antarctica are 'frozen' under Article IV of the Antarctic Treaty of 1959. Nevertheless, as this article later discusses, the ATS provides an interesting illustration of a governance regime and any stalemates regarding the fragile environmental impact of resources activities should not be connected to the failure to develop a property regime.[21]

15.2.2 Common Heritage of Mankind and the Question of Space Resources Activities

Article 11, para 1 of the Moon Agreement, which provides that the Moon and its natural resources are the 'common heritage of mankind' (CHM), is much more problematic than non-appropriation issues. It states that the parties undertake to establish an international regime to govern exploitation of natural resources; as such, exploitation becomes feasible.[22] The term 'common heritage of mankind' appears for the first time in the Moon Agreement, but it has its origin in the United Nations Law of the Sea Convention. As a result, and as the Cologne Commentary on Space Law notes, 'there is no unique feature or specific meaning of the CHM in all its manifestations'.[23] This problem of interpretation has been identified since the early stages of CHM literature.[24]

From a legal standpoint, the CHM is an evolution of the *res communis* concept. Under this view, certain areas beyond national jurisdiction may not be appropriated nor occupied by any states, because they represent a common heritage of all mankind. At the same time, such a theory calls for equality of treatment and confers on all states the right to explore, use and exploit the common area and its resources.[25] However, this approach differs from CHM, according to which any area beyond national jurisdiction should be commonly managed by all states on behalf of mankind.

---

[20] John G. Wrench, 'Non-Appropriation, No Problem: The Outer Space Treaty Is Ready for Asteroid Mining' (2019) 51 Case Western Reserve Journal of International Law 437.

[21] Ibid.

[22] Tanja Masson-Zwaan and Mahulena Hofmann, *Introduction to Space Law* (4th edn, Wolters Kluwer 2019) 34–35.

[23] S. Hobe, B. Schmidt-Tedd & K.U. Schrogl (eds), *Cologne Commentary on Space Law vol. II* (Carl Heymanns Verlag 2013), 395.

[24] E.D. Brown, 'Freedom of the High Seas Versus the Common Heritage of Mankind: Fundamental Principles in Conflict' (1983) 20 San Diego Law Review 521, Mary Victoria White, 'The Common Heritage of Mankind: An Assessment, (1982) 14 Case W. Res. J. Int'l L. 509.

[25] Fabio Tronchetti, 'Legal Aspects of Space Resource Utilization' in Frans von der Dunk and Fabio Tronchetti (eds), *Handbook of Space Law* (Edward Elgar 2015).

Unlike the theory of *res communis*, CHM is related to a common management concept of resource utilisation aligned with a pre-established international regime, such as the International Seabed Authority.

The current space law literature states that, 'while different interpretations may be put forward, the proper meaning of the CHM can only be determined in the context of its use and for the purpose of the future applicable regulatory regime'.[26] In 2008, a joint statement of state parties to the Moon Agreement determined that the CHM principle does not constitute an obstacle to space mining activities.[27] Although this was a declaration, the shift in interpretation could represent the first step to a common understanding between signatory parties based on their previous experience and future intentions.

In summary, the OST's fundamental prohibition to acquire new state territory by any means and the controversial provision of CHM fail to fully address issues such as commercial exploitation of natural resources or the settlement of permanent and self-sufficient human stations on planets. As some scholars have observed in the debate on space resource activities, there are two different interpretations possible. First, countries such as the United States and Luxembourg agree that Moon and other celestial bodies are global commons, equating such activities with the law of the High Seas, which means that entities involved in space resources activities can exercise their relative rights by obtaining authorisation from national authorities.[28]

The second interpretation posits that, since outer space belongs to all of mankind according to the OST, all available natural resources also belong to all of mankind.[29] Some countries support this principle by asserting that the Moon belongs to humanity as a whole. Thus, space activities should guarantee humanity-wide benefits such as the regime of the International Seabed Authority.[30] For opponents of commercial space mining projects, this provision means that nothing in outer space can be appropriated. This line of argumentation is considered to be further supported by the existence of the CHM principle in the Moon Agreement.

---

[26] Tanja Masson-Zwaan and Mahulena Hofmann, *Introduction to Space Law* (4th edn, Wolters Kluwer 2019) 101.

[27] Joint Statement on the benefits of adherence to the Agreement Governing Activities of States on the Moon and Other Celestial Bodies of 1979 by State Parties to that Agreement, UN Doc. A/AC.105/C.2/2008CPR.11 (2 April 2008).

[28] Frans von der Dunk, 'Who owns the moon? A space lawyer answer' (University of Nebraska-Lincoln, 20 July 2018). https://digitalcommons.unl.edu/cgi/viewcontent.cgi?article=1109&context=spacelaw.

[29] Frans von der Dunk, 'Asteroid Mining: International and National Legal Aspects' (2018) 26 Mich. St. Int' L. Rev. 83, 101.

[30] Frans von der Dunk, 'Who owns the moon? A space lawyer answer' (University of Nebraska-Lincoln, 20 July 2018). https://digitalcommons.unl.edu/cgi/viewcontent.cgi?article=1109&context=spacelaw.

15.3 Drawing Lessons from the Past, Drafting Disruptive Rules for the Future

In the age of Space 4.0, questions are starting to be raised about whether a treaty that was negotiated prior to the evolution of technology that makes settlement on another planet feasible can continue to provide an appropriate foundation for space governance of the Moon and Mars and face new challenges.[31]

The world's major powers—which are the only countries capable of sending humans to Mars—are not parties to the Moon Agreement. Furthermore, as previously mentioned, parties to the agreement undertake to establish an international regime to govern the exploitation of the Moon's natural resources; as such, exploitation is imminently feasible (Article 11, para 5). Current industry attempts to commercially use space resources make this matter even more urgent.

The same critics raised also about the OST, as many scholars view it as merely an outdated relic from the Cold War era. In this context, 'if the Treaty were repealed or interpreted to allow a free-for-all, method of allocating space property rights, this would produce an unequal allocation of resources'—especially in a manner that disadvantages developing nations, as they lack such access to space.[32]

However, the absence of a clear multilateral regime that addresses issues of space resources activities and human settlement on planetary surfaces and complements international space law and national laws stifles any private activities with positive and honourable motives. As a result, it is necessary to change the narrative and distinguish such activities from the interpretation deadlocks of the past and develop a disruptive legal approach to the establishment of an effective regime for the Moon and Mars.[33]

15.3.1 A Mars Agreement: Following the Antarctic Treaty System Model

Antarctic governance provides a useful case study of regional cooperation, beginning with the adoption of the Antarctic Treaty and followed by subsequent agreements such as the ATS. Some of its advantages are discussed below.

First, an interesting aspect is state participation. The question of state participation in the ATS has been a core issue from the beginning. Indeed, the treaty originally

---

[31] Ulrike M. Bohlmann, 'Space 4.0' in Stefano Ferretti (ed), *Space Capacity Building in the XXI Century* (Springer 2020).

[32] Abigail D. Pershing, 'Interpreting the Outer Space Treaty's Non-Appropriation Principle: Customary International Law from 1967 to Today' (2019) 44 The Yale Journal of International Law 149.

[33] Joshua Fitzmaurice and Stacey Henderson, 'On the Legality of Mars Colonisation' (2019) 40 Adelaide Law Review 841.

aimed to reconcile the interests of two distinct groups of states to promote international cooperation on the Antarctic. On the one side, there were states that made territorial claims which covered most of the continent and sometimes overlapped with each other. On the other side, the two leading states—Russia and the United States—rejected these claims and considered themselves to have a direct interest in the continent.[34] Furthermore, there were risks concerning the use of the Antarctic for non-peaceful purposes.

All of these issues were managed by the Antarctic Treaty (1959), which provided a two-tiered approach to participation in the regime. Whilst it was open to accession by any member states of the United Nations, effective participation in the decision-making process was limited to a more exclusive group of states referred to as Consultative Parties.[35] The 12 original signatory states were automatically Consultative Parties, whilst acceding states could become so based on a requirement related to the realisation of scientific activity.

By the same token, in a potential Mars Treaty System, a variable geometry approach to participation—a very common approach to economic integration procedures—could reconcile the interests of states that can and states that cannot reach Mars.[36] The differentiation between consultative and non-consultative states would contribute to this idea and ensure that only truly active participants could take full responsibility under the treaty.

Other subsequent agreements in the ATS, such as the Convention for the Conservation of Antarctic Seals, had very strict conditions for the accession of states that were not party to the original process. By contrast, conditions for participation in the Convention for the Conservation of Antarctic Marine Living Resources were less exclusive than in the Antarctic Treaty.

Another important characteristic of the ATS is the participation of non-governmental organisations, especially under the Environmental Protocol of 1991.[37] Owing to the fact that space agencies and the private space sector have jointly framed the era of Space Exploration 3.0, the Mars Treaty System could follow the same pattern.[38]

Although there may be many more key characteristics of the ATS to be examined, this analysis limits itself to the main benefits of a treaty system regime based on a variable geometry method of participation. A Martian governance regime, as Antarctica's, will evolve under similar economic conditions. During the emergence of the ATS, a major driving force was the expectation that economic activities may

[34]Costas Hadjiconstantinou, *The Legal Regime of Antarctica* (Paratiritis 1991) 28.

[35]Sebastien Duyck, 'Drawing Lessons for the Arctic Governance from the Antarctic Treaty System' (2011) 3 The Yearbook of Polar Law.

[36]Yves Bertoncini, 'Differentiated Integration and the EU: A Variable Geometry Legitimacy' (IAI, Roma, 10 March 2017).

[37]Ibid.

[38]Quoted in Joshua Fitzmaurice and Stacey Henderson, 'On the Legality of Mars Colonisation' (2019) 40 Adelaide Law Review 841.

proliferate in the near future and carry environmental risks.[39] The parties' response to the ATS was to adopt a proactive and precautionary approach to the regulation of such activities. Implemented to its full extent, the precautionary approach led to the adoption of moratoriums and regulatory frameworks guaranteeing that economic activities did not occur in the absence of legal norms.[40]

Thus, a Mars Agreement could serve as the cornerstone of a Mars Treaty System by establishing and maintaining an adequate platform for cooperation that is able to address new issues before they emerge.

15.3.2 The First Settlement on Mars. The Antarctic Stations and the ISS Model

Before examining some successful human station regimes in areas beyond national jurisdiction, it would be useful to study an older background paper on space stations published by the Office of Technology Assessment of the United States Congress. According to the paper, 'a space station could have at least four different types of legal status, making it either:

(a) a national space station under the jurisdiction and control of a single nation,
(b) a multinational space station under the joint jurisdiction and control of several nations,
(c) a multinational space station the individual modules of which are under the independent jurisdiction and control of separate nations, or
(d) an international space station under the jurisdiction and control of and international authority'.[41]

Depending on these options, the rights and liabilities of a state and its citizens could be substantially different. Assuming that the space station is solely owned and registered by a state under the terms of the Registration Convention of 1975, its legal status would be similar to that of other space objects, except for the fact that Moon stations must be open to inspection by other parties to the Moon Agreement, which could visit them after providing reasonable notice.[42]

In the case of a multinational space station under joint jurisdiction and control, there is no provision of joint registration according to the OST (Article VIII). Although the Registration Convention does not clarify which states have jurisdiction

---

[39] Yves Bertoncini, 'Differentiated Integration and the EU: A Variable Geometry Legitimacy' (IAI, Roma, 10 March 2017).

[40] Ibid.

[41] U.S. Congress, Office of Technology Assessment, Space Stations and the Law: Selected Legal Issues, Background Paper, OTA-BP-ISC-41 (Washington, DC: US Government Printing Office, August 1986).

[42] Hamilton DeSaussure, 'The Impact of Manned Space Stations on the Law of Outer Space' (1984) 21 San Diego Law Review 985, 997.

and control over unregistered objects, states must choose which will be the registering state.

According to the third type of legal status, a space station could conceivably be composed of different modules that are each owned and registered under the jurisdiction and control of separate countries. Finally, the last type of legal status seems to be an early version of the ISS model.

On Mars, human settlements could follow one of the last two abovementioned types of space stations. However, according to some scholars, the ISS model is one of the most significant examples of how the principle of international cooperation has been successfully applied.[43] The ISS Intergovernmental Agreement of 1998 is the primary legal document related to ISS activities and effectively regulates registration and jurisdiction issues as well as intellectual property rights.[44] In terms of criminal jurisdiction, the applicable criterion is the personal jurisdiction of nationality; however, this issue is much more complex,[45] as it also relates to state's responsibility and aspects of liability.[46]

After the US administration requested in February 2018 to end direct federal funding for the ISS beginning in 2025 as part of its (FY) 2019 budget request (in a move that was designed to free up funds for NASA's future projects), it will be very interesting to see how this privatisation process will work. According to a review by the NASA Office of Inspector General, NASA may need to dispose of the ISS within the next few years. This is because 'transitioning the ISS to private operation under the timetable currently envisioned, presents significant challenges in stimulating private sector interest to take on an extremely costly and complex enterprise'.[47]

The Antarctic stations model provides a successful cooperative regime based on the 1991 Madrid Protocol, which was adopted under Article IX of the Antarctic Treaty and establishes detailed principles and specific obligations in its annexes. According to a 2000 study, useful examples could also be drawn from polar stations on the former Spitsbergen (under the 1925 Norwegian Svalbard Act), which enjoys an international legal status that is similar but distinct from that of the Antarctic.[48]

---

[43] Joshua Fitzmaurice and Stacey Henderson, 'On the Legality of Mars Colonisation' (2019) 40 Adelaide Law Review 841, Carla Sharpe and Fabio Tronchetti,'Legal Aspects of Public Manned Spaceflight and Space Station Operations' in Frans von der Dunk and Fabio Tronchetti (eds), *Handbook of Space Law* (Edward Elgar 2015).

[44] Carla Sharpe and Fabio Tronchetti, 'Legal Aspects of Public Manned Spaceflight and Space Station Operations' in Frans von der Dunk and Fabio Tronchetti (eds), *Handbook of Space Law* (Edward Elgar 2015).

[45] Stephen Gorove, 'Criminal Jurisdiction in Outer Space' (1971) 6 International Lawyer 313.

[46] Sergio Marchisio, 'National Jurisdiction for Regulating Space Activities of Governmental and Non-Governmental Entities, (United Nations, Bangkok, November 2010).

[47] NASA, Office of Inspector General, NASA's Management and Utilization of the International Space Station, Report No. IG-18–021 (NASA 30 July 2018). https://oig.nasa.gov/docs/IG-18-021.pdf.

[48] Jacek Machowski, 'The Status of Stations Under the Antarctic legal regime' (2000) 21 Polish Polar Research 99.

15.3.3 A Martian Condominium?

Following the previous comments on principles of non-appropriation and CHM, the legal status of a human settlement on Mars and of Martian resources is a highly controversial issue. Traditional international space law lacks clear and internationally accepted definitions on the rules that govern space resources. Any attempt to define a specific legal framework faces controversies over the rights and prohibitions under the OST and the Moon Agreement, far from political ones.

As a result, under the planned intensity of space resources activities and the deadlock evolving from diverging interpretation, the United States and Luxembourg have adopted specific legislation that would enable them to define the conditions for the authorisation of space mining activities.[49]

The 2015 US Commercial Space Launch Competitiveness Act authorises citizens of the United States 'to possess, own, transport, use and sell the asteroid resource or space resource they obtain in accordance with the applicable law, including the international obligations' after obtaining authorisation from domestic authorities.[50] However, under this law, the United States 'does not thereby assert sovereignty or sovereign or exclusive rights or jurisdiction over, or the ownership of, any celestial body'.[51]

Similarly, the 2017 Law on the Exploration and Use of Space Resources of Luxembourg regulates space resources activities. Compared to the United States law, the beneficiaries must not necessarily be citizens of Luxembourg, but at least legal entities registered in the country.

Without any doubt, a multilateral regime for regulating space resources activities would be the best option, because it would prevent uncertainties and harmonise the application of national legislations.[52] There are examples of successfully established credibility over regulated issues, such as the ITU model of frequency spectrum coordination. Similarly, there have been proposals that the 1994 New York Agreement, which amended Part XI of UNCLOS, serve as a model to be followed.[53]

Beyond the abovementioned cases, an interesting example regarding co-administration can be found in the ATS itself: the condominium concept. For some scholars, the Antarctic Treaty appears to be a unique tool designed to resolve particular issues involved in claims to Antarctica. The treaty does not attempt to solve claims to territorial sovereignty but instead forbids national claims to the continent

---

[49]Tanja Masson-Zwaan and Mahulena Hofmann, *Introduction to Space Law* (4th edn, Wolters Kluwer 2019) 101.

[50]US Commercial Space Launch Competitiveness Act, 114 Pub. L. 90 (25 November 2015); Frans von der Dunk, 'Private Property Rights and the Public Interest in Exploration of Outer Space' (2018) 13 Biological Theory 142.

[51]Ibid.

[52]Tanja Masson-Zwaan and Mahulena Hofmann, *Introduction to Space Law* (4th edn, Wolters Kluwer 2019) 105.

[53]Fabio Tronchetti, 'Legal Aspects of Space Resource Utilization' in Frans von der Dunk and Fabio Tronchetti (eds), *Handbook of Space Law* (Edward Elgar 2015).

and creates a global commons, which serves as a framework agreement to govern the use of mineral and other resources.[54]

The condominium concept was present from the beginning of consultation talks on the establishment of a mineral natural resource regime in Antarctica. Under the 'condominium plan', the Consultative Parties would declare Antarctica to be their joint and exclusive property, with full sovereign and beneficial rights vested only in themselves. Furthermore, 'a condominium promises to reconcile conflicting territorial claims by asserting indivisible, collective sovereign rights to all of Antarctica and by distributing the proceeds of resource development among co-owners'.[55]

For some scholars, Antarctica is a de facto condominium. However, by applying such a concept to Mars, it would be necessary to adjust and further develop the term in line with international space law. In contrast, a free appropriation concept of Martian non-renewable resources could allow technologically developed states to deplete valuable resources before developing states could take a fair share. A fitting analogy for Mars' non-replenishable resources would be the resources in the international seabed rather than fish in the High Seas.

15.4 Conclusion: Designing a Mars Treaty System

The ATS has evolved into a regime that covers most existing and potential activities in Antarctica, which demonstrates that successful cooperation amongst states can be achieved and maintained despite ongoing legal disputes. Some believe that, due to the unique issues at play in Antarctica, the ATS cannot be used as a model for other territorial cases.[56] However, it is questionable whether deep seabed mining is the best model to follow, considering its complexity and limited participation. The same applies to the perspective provisions of the Moon Agreement derived from Article 11, para 5.

To some extent, under a disruptive legal attempt, a new model of the condominium on Mars aligned with existing law in the OST could serve as source of inspiration. Nevertheless, such audacity requires more liberal interpretations of OST principles. How do 'descendants' of the first human settlement conceive of shared rights to land or other resources? Which law is the condominium is aligned with? These are some of the theoretical questions that underpin the more pressing practical concerns that preoccupy those charged with creating a condominium. Although establishing a

---

[54] Joel H. Samuels, 'Condominium Arrangements in International Practice: Reviving and Abandoned Concept of Boundary Dispute Resolution (2008) 29 Michigan Journal of International Law 727.

[55] 'Thaw in International Law? Rights in Antarctica under the Law of Common Spaces' (1978) 87 The Yale Law Journal 804.

[56] Joel H. Samuels, 'Condominium Arrangements in International Practice: Reviving and Abandoned Concept of Boundary Dispute Resolution (2008) 29 Michigan Journal of International Law 727.

condominium arrangement on Mars is a theoretical question, this could serve as the basis for drafting a more complex treaty to govern a future 'New Martian Republic'.

This article intentionally left aside other critical issues such as liability and property rights, as their complexity demands a different approach. Therefore, it only examined an initial framework concerning appropriation claims and the related issue of space resource activities. For now, a more promising avenue may be to begin informal work on policy and legal proposals that could serve as blueprints and nudges for international law-making in the near future, as technology prepares to make the establishment of the first human settlement on Mars a reality.

John Tziouras is a Ph.D. candidate at Aristotle University of Thessaloniki (Greece), Faculty of Law. His research interest is in space security and space law. He obtained his LL.M. degree in International and European Legal Studies at Aristotle University of Thessaloniki and works as a special advisor for local authorities.

The manufacturer's authorised representative in the EU is Springer Nature Customer Service Centre GmbH, Europaplatz 3, 69115 Heidelberg, Germany. If you have any concerns regarding our products, please contact ProductSafety@springernature.com

Printed and bound by CPI Group (UK) Ltd, Croydon, CR0 4YY
30/09/2026
02225295-0001